Managing Wildlife Habitat on Golf Courses

Ronald G. Dodson
Audubon International

Ann Arbor Press
Chelsea, Michigan

Library of Congress Cataloging-in-Publication Data

Dodson, Ronald G.
Managing wildlife habitat on golf courses / Ronald G. Dodson.
    p.   cm.
Includes bibliographical references (p.).
    ISBN 1-57504-028-X
1. Wildlife habitat improvement. 2. Wildlife attracting. 3. Golf courses. I. Title.
QL82.D62 2000
639.9′2—dc21

99-053028

Cover photo courtesy of C. J. Elfont.

ISBN 1-57504-028-X

PRINTED IN THE UNITED STATES OF AMERICA
10 9 8 7 6 5 4 3 2

In Memoriam

My Dad

Bruce Dodson

1924–1993

He taught me the game of golf and he taught me to never give up.

Acknowledgments

Like all books, this one would not have been possible without the help and support of many individuals and organizations. I would like to extend my deepest appreciation to them.

The book was made possible through a generous grant from the PGA of America. I thank them for both their support and patience. I would especially like to thank Marty Kavanaugh for his support and encouragement, not only of my own efforts, but for his belief in the value of our programs.

I would like to thank and recognize the efforts, energy, and dedication of the hundreds and hundreds of golf course superintendents that are "walking the walk and talking the talk" of environmental stewardship as they tend to their business of managing the game of golf. Their leadership is instrumental in preserving the natural beauty and heritage of golf. I would especially like to extend my appreciation to those who contributed their experiences to this book.

The United States Golf Association has been a catalyst for environmental responsibility on golf courses. Their vision and continuing support as sponsors of the Audubon Cooperative Sanctuary Program for Golf Courses has been invaluable. USGA support has provided me with the opportunity to reach thousands of golf course managers and to encourage and promote their environmental stewardship efforts. I would like to extend special appreciation to James T. Snow, Mike Kenna, and Kimberly Erusha for their expertise, guidance, and support.

Many individuals devoted their time to review and comment on the various drafts of this publication. I appreciate all of their constructive suggestions and I feel that the end result was made much better by their advice. In particular I would like to thank Peter Stangel of the National Fish and Wildlife Foundation; Robert Lohmann, Lohmann Golf Design; Tim Hiers, Collier's Reserve Country Club; Nancy Richardson, Audubon International Signature Program; Miles (Bud) Smart, Audubon International Institute, Planning Department; Larry Woolbright, Audubon International Institute, Research Department. I would also like to thank Jean Mackay, Audubon International, Director of Education, for her contributions and insight.

I would like to thank Ann Arbor Press and Skip DeWall for agreeing to publish this text, which has been "in the works" in one way or another for a number of years, and for his valued advice and guidance.

My family has made a very special and substantial contribution to this effort. My wife, Theresa, and sons, Kelly, Eric, and Travis have supported (endured?) my late night hammering on the computer as I put these words together. More importantly, I owe them a debt of gratitude for their understanding and support in the face of the hundreds of days per year that I am away from home working on one conservation project or another in some other part of the world.

Finally, I want to give a special thanks to Mary Jack. Mary has read and re-read the words in this book many, many times. She has taken my often rambling sentences and philosophical wanderings and crafted my thoughts into coherent, focused points. This book would absolutely not have been possible without her guidance, friendship, and skills. It is as much her book as mine.

About the Author

Ronald G. Dodson is the founder and president of Audubon International and chairman of the board of the Audubon Society of New York State, Inc. He is responsible for creating the Audubon Cooperative Sanctuary System (including programs for schools, backyards, corporate properties, and golf courses) which has received international recognition. Dodson also created the Audubon Signature Cooperative Sanctuary program designed for properties in the planning and design stages of development. He also founded the Audubon International Institute for Sustainable Resource Management.

Prior to his present position, he served as a regional vice president of the National Audubon Society, executive director of the Western Kentucky Environmental Planning Agency, and as a biology instructor at both the high school and college level. Dodson holds a B.S. in Wildlife Biology from Oakland City University and an M.S. in Natural Resource Management from Indiana State University.

He has received numerous awards and recognition for his conservation efforts, including the 1985 National Environmentalist of the Year, presented by the Friends of Audubon, Inc. to commemorate the 200th anniversary of the birth of John James Audubon. On behalf of the Audubon Society of New York State, he also accepted The President's Award for Environmental Leadership by the Golf Course Superintendent's Association of America. In 1998 he was also recognized by *Landscape Management* as Person of the Year, received the New Jersey Turfgrass Environmental Award, and a Special Achievement Award from Oakland City University.

He is co-chairman of the United States Golf Association's Wildlife Links Research Committee and a member of the USGA's Environmental Research Committee and Turfgrass Research Committee. He has written articles for such publications as the *Green Section Record, Golf Course Management,* and *Golf Course News*. He also serves on the board of directors of the American National Fish and Wildlife Museum in Springfield, Missouri.

About Audubon International

Audubon International is a not-for-profit, environmental organization concentrating on conservation assistance, education, and research especially in regard to conserving and enhancing biological diversity and promoting sustainable natural resource management.

Through its programs, Audubon International provides information, guidance, and support to enhance wildlife habitat and improve environmental quality. The Audubon Cooperative Sanctuary System is comprised of membership programs for existing backyard properties, school properties, golf courses, and business properties. In addition, the Audubon Cooperative Signature Program was created for projects in the planning phases of development. Through these programs, Audubon International educates and promotes stewardship and public participation in scientifically-based land management practices.

Audubon International is one of the more than 500 Audubon Societies throughout the United States today. Each of these groups is independent and separately incorporated and each is free to establish its own goals, develop its own programs, and take positions regarding environmental issues.

Audubon International, as well as many state Audubon Societies (New York, Massachusetts, Maine, New Hampshire, New Jersey, Illinois, Rhode Island, and Connecticut) are not affiliated with National Audubon Society which has chapters in many states. The diversity of Audubon Societies is not meant to confuse the public. Rather, it serves to broaden public involvement and increase the number of approaches taken to enhance and protect the environment.

Audubon International grew out of the Audubon Society of New York State, one of the first Audubon Societies to be founded in the United States. Audubon International was created to help expand efforts for sustainable resource management throughout the United States and internationally.

Foreword

Throughout this book I have used photographs and information provided by members of the Audubon Cooperative Sanctuary Program and the Audubon Signature Program. I appreciate the time and energy that all of these golf course managers have invested in documenting their environmental efforts in words and images. These are only a few of the many, many golf course managers who demonstrate their dedication to environmental stewardship through habitat enhancement and other conservation and natural resource management initiatives.

Nothing would have brought me more pleasure than to highlight the contributions and stewardship activities of every golf course member and manager who has attained the designation of Certified Audubon Cooperative Sanctuary, or the Audubon Signature Sanctuary designation. While we cannot do justice to all of the great things that are happening on today's golf courses and all the people who have worked hard not only for the game of golf, but for the environment, we hope that this book will help to support your vision and your efforts.

If you are interested in keeping up to date concerning the stewardship efforts of golf courses involved in Audubon International programs or would like more information about our programs, I recommend that you visit the Audubon International World Wide Web site at http://www.audubonintl.org.

Contents

Managing Wildlife Habitat on Golf Courses

Ronald G. Dodson

Audubon International

Introduction

Managing Wildlife Habitat on Golf Courses was written in response to a growing interest in the management and environmental stewardship of golf courses. There is a fundamental relationship between the two. Even though the primary responsibility of golf course managers is to manage a golf course as a golf course, there is something more basic about their responsibility, and it has to do with managing land for more than just people. It is a matter of respecting the land and sharing it with the wildlife that inhabit it. From an environmentalist's perspective, it has been exciting being part of an educational process that has supported golf course managers in their effort to become environmental stewards. It has been gratifying to observe their successes as they enhance the land they manage. It is important that they take pride in the efforts they have made, but it is even more important for them to share their enthusiasm so that others will follow in their footsteps. Perhaps, if you have not already joined these golf course managers, this book will help you begin your journey.

Golf course managers who become successful in their environmental efforts seem to share several qualities: commitment, communication, cooperation, and education. It is no small challenge to meet golfers' expectations and overcome differences in opinions and attitudes regarding golf course management. It requires a commitment to managing golf course property with the best interests of the environment in mind. It requires a willingness to learn, to teach, to make compromises, and sometimes to be very patient—with people, with the land, and with yourself. It also requires a dedication to a cooperative approach. That cooperation is demonstrated through a willingness to work with other people, to share with them your perspective founded on the most important key of all—education—a willingness to learn more and to share your knowledge with others.

This book is designed to be a preliminary step toward a continuing educational process by providing a basic foundation for managing wildlife habitat on golf courses. Although it is intended primarily for golf course managers who are actively managing existing golf course properties, it may also be of interest to golfers, club managers, PGA professionals, golf course architects, or others interested in the relationship between wildlife habitat and the game of golf.

1

This is not a technical manual, nor is it a science course. Rather, it is an introduction to understanding fundamental wildlife habitat management concepts and their practical application on golf courses. Although these concepts are applicable to all types of land, golf courses present some unique challenges and opportunities.

First of all, golf is a game, and it is one that is directly linked to an economic engine. It costs money to build a golf course and it costs money to manage and maintain it. Secondly, many golf course managers are faced with the constraints of a property that may not lend itself to unlimited opportunities to manipulate the landscape for wildlife. Finally, golf is a game with its roots firmly planted in a "natural" setting, and sometimes the requirements of the game and the requirements of nature are not a perfect match.

At the outset, you must believe that there is a great deal you can do to meet these challenges. This book is designed to help you. Your primary objective is to understand your property—including its limitations and opportunities for wildlife habitat—and your second objective is to apply some basic wildlife habitat management techniques to enhance the property for which you are ultimately responsible. The book is consequently divided into sections to provide you with: 1) a sense of the history and tradition of golf, as well as of the history of the scientific and conservation movement, 2) a basic understanding of general scientific concepts, as well as what wildlife and habitat are, 3) an understanding of landscaping for wildlife and the value and benefits of naturalizing your property, and 4) some practical projects that you can undertake on your golf course. We have included photos and short case studies that we hope will exemplify a vision of golf courses

Diverse habitat and woodlands at Fowler's Mill Golf Course, Chesterland, Ohio. Courtesy of Audubon International.

as potential habitat and that will help and encourage you to follow your own path toward environmental stewardship.

This book can be read in its entirety, or used as a reference for various topics concerning golf and wildlife management. Again, although it is intended for golf course managers, others who are interested in the game of golf may also develop an appreciation for the connection between golf and wildlife conservation and land management. Although books are helpful, in order to develop a true appreciation for the game of golf and the environment in which it is played—as in other areas of life—we must stop occasionally to enjoy the sights and sounds of wildlife and the nature that surrounds us.

2 Golf Courses and Land Management

Golf Courses and Sustainability

Sustainable development is a relatively recent concept in the environmental field. The term "sustainability" means using natural resources, without depleting them, in ways that support human activity. Sustainable resource management ensures that the effects of our actions do not diminish—and may even enhance—the environmental quality of life for future generations.

Sustainable resource management means caring about and appropriately managing all of the natural resources of a piece of property, as well as assessing and managing the environmental impact on surrounding properties. It includes the "built environment" as well as the preservation and restoration of wildlife habitats. It applies to decisions made regarding golf course development and management and the subsequent positive or negative impact on wildlife, biological diversity, the environment, and the human quality of life. Sustainable development should be viewed as a process, as well as a result. The golf courses of the future should be built and managed to provide optimum playing conditions for the game, but not at the expense of the environment.

People's need for space frequently comes in the form of a variety of recreational activities, one of which is the opportunity for sport. In the United States alone, over 25 million people play the game of golf. The playing field for the game of golf takes space, and to some, quite a lot of it. So we need to focus on the decisions we make about developing and managing that space. Is the water clean and healthy? Have we maintained the native biological diversity? Did we provide economically viable opportunities for a variety of citizens to enjoy the "playing field"? Have we allowed and enhanced opportunities for wildlife to coexist with humans? And finally, have we appropriately balanced the needs for this space, and given that balance, have we used this space well?

Economically, altering the land in major ways may not only require a substantial initial investment, but a continued investment to maintain the land in ways that are inconsistent with the most basic elements of the site. Understanding the fundamental aspects of nature—including the impact of changes in land use—

will help make the siting, design, construction, and management of golf courses result in more environmentally and economically sustainable golf.

Because we hope this book will be read by a variety of people who are in the golf course industry and perhaps by others who are not, it is important to clarify that this is a book about golf course management, not golf course development. Nevertheless, we must acknowledge the different opinions surrounding golf course development in the United States today, and take the opportunity to at least say a few words about the process of golf course development and the role that golf course architects, designers, and developers can play in changing the attitude of the public and promoting a conservation ethic in the golf course industry.

Golf Course Development

It is everyone's responsibility to manage the land responsibly. It is to our advantage as humans to take care of the land, to support habitat protection and natural resource conservation, and to foster ecological restoration and habitat enhancement. Unfortunately, at various times in our history we have not been very good about attending to the human impact on the land and wildlife, either from a lack of information or from carelessness or disregard. For a variety of reasons, especially during the past two decades, the golf course industry has been singled out as a major "culprit" in land use and land management practices that seem to demonstrate a lack of environmental sensitivity. As human demand for land use increases along with demand for other natural resources, the development of new golf courses increasingly gives rise to concerns about habitat loss, as well as chemical use and water quality and use, and in some instances the public's concern is understandable.

Golf's effect on the environment is a hotly debated topic in many parts of the country. Proponents of golf claim that the design, construction, and management of golf courses has no negative environmental impact. Others say golf course development is one of the worst forms of land use ever invented. Somewhere between these two divergent points of view we will probably find the facts relevant to this debate. One way of approaching the issue of golf course development is to understand a few of the steps involved in developing, designing, and building a golf course and to understand a few of the environmental issues that come into play.

The first step in golf course development is identifying land that will accommodate a golf course. Building a course will require a change in land use. That means that the land might have been previously used for agriculture, grazing, landfills, or other forms of human use. It may be land, however, that is presently undeveloped, or that is abandoned and has been left to nature's forces. Regardless of its current use, any change in land use is usually determined, restricted, or regulated by governing officials, with an opportunity for input by the public. It is

Specimen oak trees were relocated during
golf course clearing.

imperative that the public make their concerns and opinions known at the local, state, and federal regulatory levels when changes in land use are proposed. They have a responsibility to their community to ensure that the change in land use will not be detrimental to either their community or the environment. Then, once a change of land use is sanctioned by regulatory agencies, every effort must be made to ensure that negative impacts to the land are minimized and that every effort is made to preserve and even enhance the ecological integrity of the land.

If development is to take place, a critical first step is the site selection itself. There must be a commitment to selecting a site that will maximize creative project design while providing necessary buffer areas between the development area and sensitive environmental areas. In most cases, it is the developer who chooses the land. It is, however, the architect who can and should assess and report to the developer what changes can realistically and comfortably take place on the site.

Modern-day technology and heavy construction equipment have made it possible to move, shove, cut, scrape, pile, fill, and shape nearly any site to fit the most unrealistic dream a developer may have. The fact that this approach is being exercised in many areas of the country has caused stringent laws and regulations to be created, giving rise to comprehensive environmental impact statements, public hearings, and extensive court cases. This expensive government regulatory process, coupled with the use of expensive construction equipment, has driven the cost of development to astronomical levels. It is extremely important that we move toward a more sustainable lifestyle by designing less intrusive, less expensive, and more manageable developments. The golf course architect or

A barn is left standing at Cateechee Golf Club, Hartwell, Georgia, to provide historical reference and use as possible wildlife shelter.

designer plays an essential role in decisions regarding both economics and the environment.

From an environmental perspective, deciding where to put a golf course should not be based solely on available land, but should also take into consideration the ecological components of a site, because when land is "developed," the environment is inevitably changed. The biological function of a site and all the living forms in that environment—including the plant communities and wildlife species found on a site—are all altered in some way. Even though in the case of golf course construction there may seem to be an increase in certain species after construction, most often this increase is among the more common varieties of wildlife. The rare species that depended on specific types of habitat may be lost from the site altogether. So golf course developers, designers, and architects need to carefully assess the characteristics of the land, with an eye toward integrating the needs of the wildlife and habitat and human use. In addition, golf course architects and developers should at every opportunity consider the development of golf courses on properties that have been previously impacted by human activities to one degree or another. Although the economic challenges can be great, restoring degraded land and creating viable habitat and green space can be done, and it can be done to the benefit of society and the environment.

Once a decision is made regarding where a golf course will be built, there are a number of environmental "shoulds" that golf course designers and architects need to take into consideration. To begin with, golf courses should be designed *with* the land, not *over* it. **The design of the course should flow with the land.** The ultimate goal should be a golf course that looks as if it has been there forever

Preliminary routing for Colbert Hills Golf Course,
Manhattan, Kansas. Courtesy of Audubon International.

on the first day it is opened. One way to achieve this effect is for the designer to take into consideration how wildlife inhabiting the property might use it. Not unlike humans, the most basic elements all wildlife need for life are food, cover or shelter, water, and space. Considering those elements when designing a golf course will not only help wildlife management efforts, but will also ensure greater harmony between the golf course and the land. In addition, randomly and biologically integrating those elements throughout the golf course and allowing the site biology to dictate the design of the course will ultimately maximize the environmental and economic value of the site after development.

Golf courses should be designed with a clear understanding of the eco-region in which the property is located and with an eye toward improving the land and its use as wildlife habitat. Vegetation selection is critical. Vegetation selection not only impacts wildlife management, but also impacts the economic maintenance of the completed course and environmental issues relative to pesticide use and water conservation. In order to understand the true nature of the property, it is critical to identify existing vegetation. Retaining and enhancing areas of vegetation that are already part of the natural habitat of the property makes both environmental and economic good sense. Such areas will not only continue sustaining wildlife already inhabiting the property, but will contribute to lower maintenance costs by allowing self-sustaining vegetation in out-of-play areas.

Native plants to enhance existing habitat should be selected. Many times plant selections are made for aesthetic reasons, which then become poor economic reasons. Rather than selecting plants that have substantial water or chemi-

Dr. Bud Smart assessing turfgrass selection at Stevinson Ranch, Savanah Course, Stevinson, California.

cal requirements for maintenance and survival, choosing plants that are native to a particular area will not only contribute to wildlife, but will also lower maintenance costs.

Turfgrass should be selected that is geographically compatible and that will not require excessive amounts of water and chemicals to keep it in good playing condition. Species and cultivars of turfgrasses should have good tolerance to environmental stresses—such as heat, cold, drought, and shade—and be less susceptible to major disease and insect problems.

There are a number of principles for maximizing biological diversity and habitat restoration that golf designers, architects, developers, and managers should learn and understand if they are committed to making effective environmental decisions about golf courses. Because of the increasing fragmentation of the landscape, one of the most important of these is to preserve as many large blocks of habitat as possible, and when the design is finalized, those blocks of habitat should be connected by habitat corridors. Every golf course architect, designer, or developer should not only be familiar with the following principles, but should be dedicated to incorporating them in any golf course design.

Principles for Maximizing Biological Diversity and Habitat Restoration

- Large areas of natural habitat sustain more species than small areas.
- If no large areas are possible, many small areas on a property will help sustain regional diversity.
- The shape of a natural habitat is as important as the size.
- Fragmentation of habitats reduces diversity.
- Isolated patches of habitat (i.e., areas not connected to other natural habitats) sustain fewer species than closely associated patches.
- Patches of habitat connected to each other by corridors sustain more species than patches not connected.
- A mixture of habitat types in a property sustains more species than a property that has only one or two types of habitats.
- Transition zones between habitat types occur in nature. In a grass-shrub-forest progression comprised entirely of shrubs are the transition zone.
- Full restoration with native plants sustains the most diversity.
- Increasing vertical structure (i.e., understory to canopy) in a habitat increases diversity.
- The higher the diversity of plants, the more food is available year-round.
- Species survival depends on maintaining minimum population levels.
- Low-intensity land management sustains more species than high-intensity management.

Reference: *Landscape Restoration Handbook*, Donald Harker et al., Lewis Publishers, Boca Raton, FL, 1993.

Although it would be wonderful if there were one recipe for designing an "environmentally friendly" golf course, there just isn't. There is a tremendous diversity of plants, animals, temperature, soils, and water sources not only across the United States, but within one state, and even within a single parcel of land. Therefore it takes time, dedication, and patience by people who are committed to understanding their site to help that property be the best that it can be, not merely as a golf course, but also as habitat.

All species of wildlife depend upon habitat for their survival. A golf course that is appropriately designed, taking into consideration the wildlife and habitat use of the site, and that is subsequently managed not only as a golf course but also as habitat, can provide the basic needs of a variety of wildlife species. In addition, wildlife, natural habitat, and the game of golf are historically linked. The opportunity to see and hear wildlife on the golf course is an important part of the game. Golfers certainly enjoy the challenge of shooting their best scores. However, as we become an increasingly urban society, the golf course also provides an important way to "connect" with nature and to enjoy the outdoor environment.

Constructed wetlands at Breckenridge Golf Club,
Breckenridge, Colorado. Courtesy of Audubon International.

Golf Course Management

Once a golf course is built, the public's concern focuses on intensive golf course management practices and the negative impact of chemical use and water quality and use. Golf course managers have the opportunity to change this perception, and many of them have risen to the challenge. They have made the effort to identify areas that are managed as turfgrass and turned them into more productive wildlife habitat by using native plants appropriate to the region and through natural landscaping efforts. Some have even been involved in habitat restoration or ecological restoration projects.

On the other hand, many golfers and golf course managers think of a golf course only from the perspective of tees, fairways, and greens. That is because the game is played there—or at least it is supposed to be. The errant slice or hook often ends up in the rough, in the water, or in other "hazards" of the game. From a wildlife management perspective, it is these "out-of-play" areas that are extremely important.

Undeniably, the primary function of the course is for playing golf, and this is the economic engine that runs any golf facility. Therefore, issues like safety, speed of play, lost golf balls, and golfer expectations are very important. Our intent is not to create thick, impenetrable habitat exactly where all golfers have a tendency to hit their shots, nor to encourage wildlife to run rampant all over the golf course. But why maintain golfing conditions and golfing "habitat" where no golfer ever goes when we could create valuable wildlife habitat there instead? Many golf

Naturalized area at Hominy Hill Golf Course, Colts Neck, New Jersey.
Courtesy of Audubon International.

courses have saved thousands of dollars in management expenses every year just by employing this simple approach.

A habitat enhancement and natural landscaping approach not only benefits wildlife, but over time it will save the golf course money that can be reinvested in habitat conservation projects that will ultimately benefit the environmental image of the golf course industry. In other words, if people in the golf course industry want golf to continue as a growing and thriving part of our economy and of our developed landscape, then we need to get back to the roots of the game. We need to enhance the "natural" landscape of golf courses and increase its value to wildlife not only to preserve the history of the game of golf, but to ensure its future.

Developing a balanced perspective about golf course development and management and the impact on the environment is the beginning of our effort to preserve the future of golf as well as of our natural landscapes. Through the next several chapters, we will take a brief look at the history of golf, the evolution of the conservation and scientific movement in the United States, and the origins of field biology. At the end of this journey, we hope you will have a better sense of the relationship among events and the development of a conservation ethic. Our environmental progress has created a unique opportunity for today's golf course managers to take the lead in a positive environmental approach to managing the land for wildlife, humans, and the game of golf.

The Evolution of the Game of Golf and of the Conservation Movement

In order to develop a foundation for wildlife management practices, it is important to look back at history, and in doing so to develop a broader understanding of golf and the evolution of the conservation movement. They are not separate events, but rather a progression of many events over long periods of time that have brought us to the present, and that will help us determine where we want to be in the future and what we have to do to get there. Again, the purpose is to provide background information, some of which you may already know, but some of which we hope will give you a broader understanding of the connection between the traditions of golf and wildlife habitat conservation and management.

A Short History of the Game of Golf

The combination of a tool, an object, and a target can be traced back to our earliest knowledge of humans. At some point in our evolution, this combination provided opportunities for competition and recreation. It is the foundation for hockey (a stick, a puck, and a net), for tennis (a racket, a ball, and a competitor), for archery (a bow, an arrow, and a target), and for golf (a club, a ball, and a hole).

In golf, the object to be struck has been a stone, a wooden ball, a leather-covered and feather-stuffed ball, and a ball made of combinations of plastic and rubber. The hitting implement has been made of wood, wood and iron, aluminum, and fiberglass and titanium. The targets have been a pole, a door, a tree, and a hole. The aim has always been to get the object to its destination in the least number of strokes.

There is some written evidence and some artwork that suggest that the Dutch may have created a forerunner of modern golf, which was then refined by the Scots. On the other hand, some believe that the Dutch game simply resembled a Scottish game. The Dutch game, kolfspel, involved hitting a rock with a crooked stick or club called a "kolven." The Dutch version was played on ice with a rock, which was hit toward a post, or on grass, where a hole was the target.

15

The traditions of modern golf, however, have been traced to St. Andrews, Scotland. The city of St. Andrews developed a port near the estuary of the river Eden. The winds and seas of the area had shaped and reshaped the land and the coast for years. The land to the north of St. Andrews, open and windblown, was reclaimed from the sea by the natural processes associated with wave and wind action. This sandy land was referred to as "linksland" because it linked the land suitable for cultivation to the sea. The linksland that was reclaimed from the sea was used in early days for grazing sheep, which thrived on the short, sea-washed grasses. The windblown holes and scattered bushes provided shelter for both the shepherd and the sheep.

In the year 1123, the area of the links was granted to the borough of St. Andrews by a charter signed by King David. At that time, there was no reference to golf, and it seems likely that the linksland was used as a source of peat for fuel, turf for roofs, and pasture for rabbits (which subsequently fed people). In the early 1400s, human intervention began. Townspeople began planting sand-binding grasses like marram and sea lyme, which assisted in the formation of sand dunes.

It is not clear when people actually began playing golf on the linksland. We do know that in 1451 golf was banned in Scotland by James II because he felt the game was interfering with time that should have been devoted to honing archery skills for war. James III repeated the prohibition in 1471, and so did James IV in 1492. Following a peace treaty between England and Scotland in 1502, however, the Scots were allowed to resume their golfing activities.

The first mention of golfing rights is in the charter granted by Archbishop John Hamilton in 1552, in which he confirmed, ratified, and approved the right of the community, among other things, in "playing at goff, futball, schuteing at all gamis with all uther manner of paystyme as ever thai pleis." (*playing at golf, football, shooting at all games with all other manner of pastime as ever they please.*") James VI finally settled the golf vs. archery debate when in 1603 he succeeded to the English throne and moved to London with his court, which included a number of golfers, who practiced their game on the Black Heath Common.

Over the years, various charters were confirmed and established on the linksland. The borough of St. Andrews was permitted to lease part of the links, but they were always careful to include safeguards for the protection of the golfing areas. In September of 1726, for example, William Gib, Deacon of the Baxters, was allowed to put his black and white rabbits on the links, provided that "the links are not to be spoiled where the golfing is used."

As popularity grew, so did the effort to refine and standardize the game as well as the playing surface. As golf became more organized, golf disputes were first settled by senior players on the "course." Ultimately, however, that gave way to the need for a standard code of rules. In 1744 the Honorable Company of Edinburgh Golfers standardized rules and standards with the "Thirteen Articles and Laws in Playing The Golf." They provided a silver club to be competed for

Old Course at St. Andrews, Scotland.
Courtesy of Audubon International.

annually, and the winner of the club became the Captain of Golf and was the final arbiter of all disputes. In 1754 the Royal & Ancient Golf Club (R&A) of St. Andrews, Scotland, was established, and as the game became more standardized, it spread from Scotland to the rest of the British Isles and then to the European continent, North America, and elsewhere.

The oldest golf club in the United States for which there is certain evidence is the South Carolina Golf Club, which was established by British planters in Charleston in 1789. It was closed in 1812, but has been rechartered in recent times. By the 1830s, golf was well established in the United States, and by the 1840s many golf courses had a professional who did everything from making and repairing clubs and balls to tending the greens and providing instruction. In 1886 the first permanent golf club was established in the United States by John G. Reid, often credited as "the Father of American Golf." Originally from St. Andrews, Scotland, he called his club the St. Andrew's Golf Club, which still exists in Hastings-on-Hudson in New York State. In 1894 the United States Golf Association was created, which along with the Royal & Ancient Golf Club is responsible for making decisions about the rules of golf and organizing competitions.

In 1900 the United States had over one thousand golf courses. By 1910 fairways were mowed using horses pulling single mowers. In 1914 Charles Worthington

USGA Golf House, Far Hills, New Jersey.

introduced a three-unit horse-drawn mower, and in 1919 he built a golf course tractor. By the end of World War I, the use of multi-blade steel mowers on greens was universal. The original links courses were shaped by wind and water. Natural plants and grasses comprised the rough and the variety and diversity of vegetation and topography added beauty and challenge to the game. By the early 1900s, however, the term "golf course architecture" became more commonplace.

The courses that were built during the Classic Period (1900–1930) both in Britain and America have withstood the test of time. In Britain, Herbert Fowler and Tom Simpson, Abercomby and Croome, Harry S. Colt, Alister MacKenzie, and Charles Alison all led the way toward modernizing golf course design and management. In America the trend was set again by Colt, MacKenzie, and Alison, who had moved from Britain to develop golf courses in the United States. Individuals such as Charles Blair Macdonald, Seth Raynor, Charles Banks, Donald Ross, and A.W. Tillinghast had a major impact on golf course architecture, as did George C. Thomas Jr., and George Crump, who designed Pine Valley in New Jersey, Jack Neville, who designed Pebble Beach in California, and William and Henry Fownes, who designed Oakmont in Pennsylvania.

The Great Depression in the 1930s, however, interrupted the continued expansion of the golf course industry. After World War II, the Modern Period of golf course architecture (1950 to the present) began with the popular golf course architect Robert Trent Jones Sr., known as "the Dean of American Golf Course Architects." He is most well known for his redesign of Oakland Hills for the 1951 U.S. Open Golf Championship, as well as designing or remodeling over five hundred golf courses worldwide.

So, in the beginning, golf and nature coexisted because courses were "laid out" on lands as they were found. "Hazards" were really the hazards of nature—tall grasses, sandy dunes, animals, and other people strolling across the land. As time went on, manipulation of the land began, but was restricted to work done by hand and mule, until the rise of the Industrial Revolution, the creation of powerful machines, and the incredible expansion of technology since the beginning of the twentieth century.

Over time, technology has had a tremendous impact on the design of golf courses. Older courses have been remodeled and new hazards ranging from innovative bunker design to island greens have all been made possible by modern technology. Unlike the original links courses that were designed by nature, new courses are designed by professional golf course architects and built by construction crews and a cadre of consultants and specialists. Golf course construction and maintenance techniques have become increasingly sophisticated.

Technology has also impacted the visibility of the game of golf, the popularity of professional golfers, and golfers' aesthetic expectations. Televised golf has played a major role in golf course design. That is, some golf courses are designed not only to be challenging to the players, but exciting for spectators and television viewers. Consequently, the perspective of the television viewer and the physical impact of those attending the tournament are also considerations in the design. The visual impact of the golf course, however, is a prime consideration of televised golf and influences golfers' perception of what golf courses "should" look like.

It is the perspective of the viewers—most of whom are golfers themselves—that has influenced the demand for a specific vision of how golf courses are supposed to look. Unfortunately, the heavily manicured, emerald green golf courses so often shown on television also symbolize the public's concern about the overuse of chemicals and natural resources to produce these "perfect" greens. That concern is even more vocally expressed by the resistance to construction of new golf courses, especially in light of impacts on habitat, wildlife, and natural resources. But that concern did not come overnight. It was the result of several hundreds of years of human activity, during which exploration of new land took place, knowledge of science grew, and industrialization expanded from Europe to North America.

The Conservation Movement in the United States

As we've seen with the beginnings of golf, there was a sense of connection between recreation and the natural landscape. This connection was based on a fundamental relationship between humans and their environment. Prior to the 1600s, human survival was dependent on the land and natural resources—food and water for nourishment, cover or shelter for protection and reproduction, and adequate space in which to live, grow, and hunt. Agriculture and hunting were the primary activities that allowed for this survival, and they required people to have a fundamental understanding of the plants and animals that supplied the basic needs for human survival. There were few opportunities available for either recreation or formal education.

The growth of the Industrial Revolution during the 1700s and 1800s in Western Europe precipitated changes in every aspect of life. The printing press allowed for greater access to information, which in turn fostered the exchange of information, and increased discoveries in science and technology. People developed a greater ability to change the structure of the land and alter the environment. During that time, power-driven machines and more efficient tools were developed, factories were created, business and commerce grew, and opportunities for exploration increased. When Europeans came to America, they brought with them the belief that if they were to survive the great unknown wilderness that faced them, they needed to overcome the "foreign" environment. They cleared forests to make way for farms, killed wild animals not merely for their own survival but for trade, and acted with the belief that there was an inexhaustible supply of water, trees, plants, and animals.

The pioneers of America were unaware of the ultimate negative impact caused by their immense energy to "settle" the land. Their attitudes were a product of the times and the circumstances. There was neither time nor educational resources that could help people understand anything other than their own immediate needs. The prevalent attitude through the latter part of the 1800s was that the natural resources of America were inexhaustible. Traders, trappers, trail blazers, and travelers gave every impression that the New World went on forever, that there would always be more land, more forest, more wildlife, and more minerals. It was a time of rapid population growth, industrialization, and urbanization—especially east of the Mississippi—and increasing western migration.

By the end of the 1800s, however, the myth of "inexhaustibleness" came to an end. The last mountain had been crossed, the last great forest had been introduced to the ax, the last individual of several species of wildlife had been killed by hunting or for sport, or perished in the isolated confinement of a zoo.

During the early colonial and pioneering years of America, there was time for only the most basic education. Little was known about the natural history of America, and little information was available about the native plants and wildlife. There was little understanding of the uniqueness of the wildlife or other natural resources. Until the settlers had the time and the economic and educational foundation to inventory and document it—and the desire to look more closely at what comprised the American "habitat"—there would be no great understanding of what and how much we had used up. Conservation would only become an issue when we came to realize that there were limited resources and that we had used up a great deal that could not be replaced.

After the slow growth of educational institutions and cultural centers in the United States during the late 1800s, and the increasing growth of government resources in the early 1900s, it became increasingly apparent to a number of organizations that we were at risk of losing forever part of our natural heritage that we could not replace. At that time, it became increasingly evident to professional foresters that our forests were on the verge of complete destruction. Members of the newly formed American Ornithologists Union were uncomfortably aware that the passenger pigeon and Labrador duck had passed beyond recall and that many other game birds and mammals might soon follow them down the road to oblivion.

It was obvious to those who cared about nature that it was time to come together to develop a plan and to implement a strategy to conserve wildlife and the habitats that supported it. In the late 1890s, the first Audubon Society, led by George Bird Grinnell and others, was engaged in a fierce battle to protect endangered species of birds such as egrets from complete extinction, and foresters were urging the "long look ahead" and sustained yield forestry.

In government, the few men who were in a position to realize the problem were fighting the abuses involved in giving away the public domain—those millions of acres of public land distributed under the Homestead Act of 1862, the Timber and Stone Act of 1878, and other related legislation. The Friends of Conservation had succeeded in passing through Congress in 1891 a bill to permit creation of forest reserves within the nationally owned forests. A commission appointed by Grover Cleveland in 1896 had examined the nation's forest resources and made recommendations on handling them, but Cleveland, a firm supporter of conservation measures, left office the next year with most of the nation's resources still in danger.

The turn of the century brought a change in attitude regarding protecting and preserving America's natural resources. In 1901, Theodore Roosevelt, an outdoorsman, first president of the Audubon Society of New York State, a hunter, and a competent field biologist as well as politician, became president of the United States. With strong allies in Congress and government, conservation acts, began to pass into the law books one after another. Acts forbidding feather traffic and authorizing setting aside vast areas for scientific or recreational purposes were among the many acts passed affecting natural resources of the nation.

By the 1930s the concept of game management came to the forefront. There was a greater application of scientific knowledge and techniques to maintain, limit, or increase wildlife populations and habitats. Legislation was passed to protect public lands and wildlife, and a number of wildlife agencies were established to acquire land, perform wildlife research, and conduct programs and training. The public applied more pressure through conservation organizations to assist in preserving and protecting wildlife species and their habitat.

By the mid-1900s, public concern for wildlife and habitat protection broadened to include pollution, pesticide use, and plants. Our increasing awareness and understanding over the past four hundred years has brought us to the realization that the loss of habitat—including wetlands, forest, and desert—is the pri-

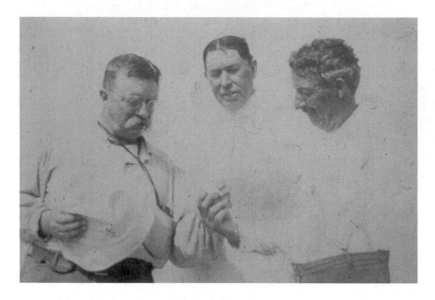

Theodore Roosevelt (left) in the field.

mary reason for the continuing decline in wildlife populations. Loss of habitat for developed land like housing, roads, shopping centers, and landfills encroaches on the limited amount of food, water, cover or shelter, and space required for wildlife to reproduce and raise their young.

In the United States, we have been blessed with an abundance of natural resources. The very foundation of American history lies in what have seemed to be unlimited natural resources and it has given us a false sense of security. As the world's population increases and the standard of living rises, we too will face increasing competition for limited supplies of many natural resources. We have begun to understand that the only way to ensure that these natural resources will be available for future generations is to find ways to conserve and protect those resources.

The concept of conservation is relatively new, and we seem to be only gradually coming to it by looking at our past and understanding the choices and decisions we made about how we used the land. Education and awareness are sometimes slow processes, and they influence our ability to act in ways that serve our own best interests or the interests of generations yet to come. In our recent history, the greatest influence has been that of the scientific world. It has allowed us the opportunity to explore, study, and hopefully bring a greater understanding to our world. Our understanding of the evolution of the sciences and our knowledge of the plants and animals of America has also played an important role in our current understanding of the need to conserve and protect natural resources.

John James Audubon.
Courtesy of Audubon
International.

John James Audubon was born in 1785 in Haiti. Although many people are familiar with the name "Audubon," they know little about the life of the man. John James was the son of a French naval officer and spent the greater part of his youth in France, where he had the opportunity to study art with the famous painter Jacques-Louis David. When he was eighteen years old, he was sent to America to help supervise his father's plantation in Pennsylvania. There he met and married Lucy Bakewell. In 1808 they moved to Kentucky, where Audubon found new and abundant wildlife to paint and study. After several failed commercial ventures, Audubon finally declared bankruptcy and was forced to make a living by doing what he had done for pleasure—paint the birds of North America.

Audubon traveled extensively throughout the United States to gather specimens. His drawings were the first of their kind in that they portrayed the subjects in such lifelike positions and in daily activities, including adult birds feeding their young, fighting predators, killing, perching, and flying. There were few reference materials available. What he knew about birds, he learned for himself by studying them.

Unable to find a willing engraver in the United States, Audubon sailed to England to find an engraver for these magnificent drawings. These paintings became the first ever to be published life-size. They were untrimmed, measuring 39-1/2 inches by 29-1/2 inches, and were sold in sets referred to as "Double Elephant Folios." Each set consisted of four volumes and contained 435 hand-colored engravings showing 1,065 life-size figures of 489 distinct species. The set was called *The Birds of America.* Upon completion, the four-volume set would have sold in the United States for $1,000. Never before had such a project been attempted, and even today it remains unique and unequaled in scope—a true monument to John James Audubon. Today modern ornithologists find the color and form of his birds to be true to life.

Plate No. 265.
Audubon Woodpecker.
Courtesy of Audubon
International.

John James Audubon lived in an era during which hunting and killing large numbers of animals was an accepted way of life. Hunting was for sport as well as a method of obtaining food. However, during the latter years of Audubon's life, he began to understand that the extensive ravages of nature could not continue or else there would be no deer or fish or bison for future generations. He predicted that in a century neither swamp, river, nor mountain would be seen as he had seen it.

The History of the Audubon Movement

During the late 1880s, in response to the vast destruction of birds in the name of fashion, George Bird Grinnell, editor of *Forest and Stream Magazine*, founded the first Audubon Society. The fanciful style of wearing bird feathers in hats and bird wings on coats nearly caused the extinction of several species. To change this fashion trend, Grinnell used his magazine to organize a national bird protection organization. As a boy Grinnell was tutored by Lucy Audubon, widow of famed bird artist, John James Audubon, and he was greatly influenced by Audubon's passion for birds. Because he believed that

Feathers for Fashion.

Audubon "had done more to teach Americans about birds of their own land than any other who lived," Grinnell thought that "Audubon" would be a fitting name for the movement.

Grinnell also felt the best way to create change was to encourage the collective action of individuals. He urged women to pressure the fashion industry by signing pledge cards that promised they would refrain from wearing bird feathers. Men promised to shoot birds only for consumption. In order to have the greatest impact and to reach as many people as possible, he helped form small, grassroots groups dedicated to bird preservation throughout New York and other states.

Massachusetts Audubon was formed in 1896, followed by New York State Audubon in 1897. During the next five years, thirty-five Audubon Societies were incorporated, which later joined to form a loose coalition of independent state groups, The National Association of Audubon Societies. This organization, now known as the Audubon Alliance, is still comprised of independent state Audubon Societies.

As with most social and political movements, there were changes in direction, focus, and structure of these original Audubon Societies over the years. In the 1940s, a small group of individuals decided to form a separate organization that would focus on issues they felt were beyond the scope of state Audubon Societies. This organization became the National Audubon Society.

Audubon Societies Today

Today, there are more than 500 Audubon Societies in the United States. Each of these groups is independent, separately incorporated, and guided by its own board of directors. Each independent Audubon organization is free to establish its own goals, develop its own programs, and take positions regarding environmental issues. The state Audubon Societies of New York, Massachusetts, Maine, New Hampshire, New Jersey, Illinois, Rhode Island, Connecticut, the Audubon Naturalist Society of the Mid-Atlantic States, as well as Audubon International, are not affiliated with the National Audubon Society. The diversity of Audubon Societies is not meant to confuse the public. Rather, it serves to broaden public involvement and increase the number of approaches taken to enhance and protect the environment.

The Evolution of the Scientific Movement and the Origins of Field Biology

The Evolution of the Scientific Movement

People have always been interested in their environment, in the world that surrounds them. Prehistoric cave paintings represent humans' earliest interest in documenting wildlife and plant life surrounding them. As a result, we know that early humans used plants for medicines and food, animal pelts for clothing, bones for tools, and meat for food, and that this connection with the land and natural resources is part of our heritage. Aside from the utilitarian value of natural resources, there has been an ever increasing interest in and curiosity about the world around us.

Unquestionably, when Europeans faced a new world—a new environment—survival was their primary focus. During the 1600s, early settlers in America concentrated on figuring out how to live in the new climate and find food and create shelter. Staying alive took nearly all the time and energy they had. Opportunities for recreation or the study of nature were limited.

The pioneers, however, whether hunters, trappers, or farmers, had a vested interest in understanding the wild animals and vegetation they encountered, but few of them were able to write and even fewer understood the value of recording the knowledge they had accumulated. There were, however, European travelers who came to visit America who took notes about their observations. Travel books or the occasional notes or journal of a missionary, doctor, or other educated person described samples of interesting plants and animals, but were not necessarily reliable "scientific" information. Nevertheless, by the end of the 1600s, although a few semiscientific works on biology had been published in New England, very little information about American plants and animals had been published.

At the beginning of the 1700s, however, increasing numbers of Europeans with scientific interests traveled to America. Some came to settle, but others came for the express purpose of discovering and collecting unique samples of American wilderness. Much of the information about the status of American wildlife and plants at that time has been gathered from the books of these travelers. During the 1700s, the entire eastern coastline of America was being settled by the French,

Spanish, Dutch, and English, and they brought with them a wide range of education, professions, and financial resources.

By the beginning of the 1800s, exploratory research in the eastern United States was gaining popularity and support. Although the numbers of "researchers" was small, societies began to spring up in major cities and attracted new workers to the field. In 1812 the Academy of Natural Sciences of Philadelphia was formed. It became—and remains today—an active force in the study of natural history. In 1817, the Lyceum of Natural History of New York was founded. This organization still exists as the New York Academy of Sciences. Boston followed in 1830 with the formation of the Boston Society of Natural History. In 1863, the National Academy of Sciences was established in Washington, D.C.

Although these few societies did a great deal to advance the study of science, nature, and field biology, this period included a group of what were probably the most ill-assorted and eccentric geniuses ever assembled in a single country at one time. Most of them came from abroad—a French dancing master turned trader (John James Audubon); an eccentric, cosmopolitan of French ancestry born in Turkey and trained in Sicily (C.S. Rafinesque); and a poor Scottish weaver and poet out to make his fortune in the New World (Alexander Wilson). This group of early field biologists spanned an incredible range of interests in natural history. They were active from the East Coast to the Mississippi, and occasionally joined expeditions to the Rockies or the West Coast, collecting, observing, and describing as they went.

Until this time, most scientific efforts were privately funded and independently supported, and entomology (the study of insects) was not part of mainstream research study. The lack of focus on entomology and the government's lack of active participation lasted until the 1870s, when an outbreak of migratory locusts brought about the establishment of the National Entomological Commission. By 1895, the Commission had become the Department of Agriculture's Division of Entomology doing extensive fieldwork all over the country. These pioneer workers in governmental biology were so successful that in a short time state entomologists, botanists, and zoologists began to appear all over the country. Interestingly, it was the Department of Agriculture that first housed—and in a sense created—the Green Section of the United States Golf Association, which continues the tradition of connecting research and practical application in the field.

In 1878, the states of New Hampshire and California formed state game commissions, and similar organizations soon appeared in other states. Meanwhile, the United States Fish Commission had been formed. The trend continued with the establishment of a Division of Economic Ornithology and Mammology in the Department of Agriculture in 1885. This division later became the Bureau of Biological Survey, which was combined with the Bureau of Fisheries in 1940 to become the present Fish and Wildlife Service. State experimental stations began to appear as early as 1875, and the idea soon spread to other states.

The work of colleges and private individuals continued to progress in major cities and to expand westward. During the mid-1800s, a number of land grant and agricultural colleges were created, and they began to teach more and better biology. The publication of Darwin's *Origin of Species by Means of Natural Selection* in 1859 gave rise to biological study among both proponents and opponents of the theory. Manuals of biology began to appear although most were too technical for the layman. In 1863 the University of Massachusetts at Amherst was founded as the Massachusetts Agricultural College. Purdue University, Cornell University, the University of Maine, and the University of Kentucky were founded two years later.

By the early 1900s, biology was becoming a field so vast and compartmentalized that the early days of the untrained naturalist—the wanderer with a simple interest in recording the wildlife around him—seemed to be over. Biology in many colleges was confined to the laboratory, where great strides were made in comparative anatomy, embryology, genetics, and evolution.

Today, the study of biology is still most fundamentally the study of life, and as living beings ourselves we are naturally curious about the life that surrounds us. Over time academic disciplines evolved to focus on the variety of aspects of biology. Disciplines include the study of taxonomy (grouping plants and animals by their similar characteristics); the study of ecology (how particular organisms or species interact with each other and with their environment); the study of ethology (animal behavior); the study of anatomy (the structure of animals and plants); and the study of genetics (heredity). Regardless of our special interests, the study or understanding of life has almost universal appeal. The natural history of America emerged through the basic interest of people in their natural surroundings. This same interest forms the basis for field biology and, ultimately, wildlife and habitat management.

Field Biology

Scientific disciplines are dependent on two main approaches that may be used in different degrees: experimental and observational. The purely experimental approach depends on whether the experiment can be replicated; that is, whether you can repeat the same experiment and achieve the same results more than once. The experimental approach requires that you control certain variables to make sure you can do it the same way many times and that the results cannot be attributed to anything else.

Field biology is the study of life under natural conditions or in its natural habitat. Unlike other strictly experimental disciplines, field biology is primarily observational, and secondarily experimental. Instead of removing organisms from their native environment to study them in the laboratory, field biologists go where the organisms are to observe them. If they cannot make the necessary observations or perform the desired experiments under natural conditions, they may try to duplicate natural conditions as nearly as they can in the laboratory where certain factors can be controlled.

Replication is harder to achieve in the field than in a laboratory because it is difficult to control all the natural variables. Yet the field biologist's aims are no different from those of his fellow scientists. Perhaps the field biologist learns more through observation and less through controlled experiments and the experimental scientist depends more on probability and statistical analysis of data. But in the end, the field biologist's job is to discover the answers to questions about plants and animals and their natural habitat and to interpret the importance of these findings relative to human beings.

Scientists in any field are trained to reach their goal by a definite method of operation—a series of steps that allows them to identify problems, explore possible explanations, and reach conclusions. The **scientific method** is an approach that can be applied to any problem for which we seek a solution, including experiments in a laboratory as well as managing habitat and wildlife.

The Scientific Method		
Steps	**Process**	**Description**
1	Formulating the problem	Posing a question for which the answer is unknown. Identifying the question and formulating the problem are often the most difficult steps in a study.
2	Accumulating facts (data)	The process of collecting information from literature, research, and field observations and assembling it in a way that addresses the question that has been formulated.
3	Formulating a hypothesis	On the basis of the data that has been gathered, a possible answer to the problem may be suggested.
4	Testing the hypothesis	The answer is tested by additional experiments, observations, and fact checking that may bear on the suggested answer.

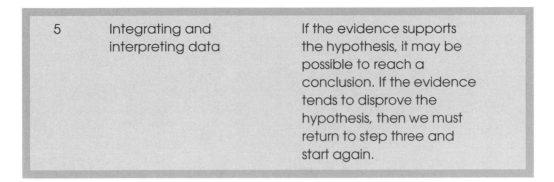

| 5 | Integrating and interpreting data | If the evidence supports the hypothesis, it may be possible to reach a conclusion. If the evidence tends to disprove the hypothesis, then we must return to step three and start again. |

A hypothesis may become a "scientific law" after it has been adequately tested and the evidence consistently supports the hypothesis. However, not all important scientific studies are made in exactly this manner. Luck may enter the picture, and the scientist may arrive at a conclusion without even formulating a question. Sometimes the answer to an entirely different question may emerge during the process of an investigation. Nevertheless, this is a simple way to explain how scientists work and the importance of setting up a process by which one scientific study can be repeated to verify the results.

This is the basic foundation of any problem-solving approach. It is a means by which we ask questions, pose answers, try out different approaches, and evaluate results. It can be a tool used by a golf course manager to understand how to deal with conflicts between wildlife and people, or it can be more a rigorous tool to study how wildlife functions in relation to its environment.

Ecology

Ecology is the more sophisticated offspring of the old "natural history." It is the study of organisms in relation to their environment. In the science of ecology, the field biologist has always played a leading role. Only by careful observation and study of nature and by having intimate knowledge of lives and habits of wild organisms can ecological data be properly gathered and correctly interpreted. The problems of forest management, range management, wildlife management, or golf course management depend for their solutions upon a broad ecological understanding of organisms in nature.

As the world's population increases and millions of people demand higher standards of living than their ancestors knew, the problems of wildlife and resource management take on added importance. The use of the world's resources for the highest and best use, for the greatest number of people and other living creatures, and for the longest length of time has become an international problem of the utmost urgency.

Today we manage our environment as never before in history, and if we want to manage it wisely, we must base our management on a sound foundation of ecological knowledge. Such knowledge must largely be gained by field study. Our ancestors, in their understandable haste to clear the land and vanquish the wilderness, had little concern for careful use of natural resources. They had little knowledge of ecological processes, and if they thought about it at all, they concluded that our resources were inexhaustible. The myth of "inexhaustibleness" still lingers in the American mind. Plants and animals become extinct, topsoil washes down rivers, water tables continue to fall, and valuable forest habitat continues to diminish in the face of human endeavors. With increasing land utilization, for subdivisions, shopping malls, and roads, many of our native wildlife species are facing serious population decline.

A wide variety of wildlife inhabits nearly every golf course (including insects, reptiles, amphibians, mammals, and birds). Some are common, everyday wildlife found in large numbers nearly everywhere. Some are endangered or threatened species. Since wildlife and golf courses coexist, it is up to golf course managers to ensure that land management practices do not negatively impact wildlife or the environment generally. This may seem a heavy burden given that golf course managers were hired to maintain the golf course for golfers. Nevertheless, whether they know it or not, to some extent golf course managers practice field biology and wildlife management every day.

Many golf course managers began their career because of their love and enjoyment of the outdoors. It is true that their primary responsibility is to manage a golf course, but it is virtually impossible to do that without being aware of the wildlife that shares the golf course, or of the variety and diversity of the land they manage. Over the past several years, as the environmental movement has made more of an impact on golf courses, the number of amateur field biologists has quietly grown. More golf course managers have become increasingly interested in learning more about the land and the wildlife that surrounds them.

Because golf course management may positively or negatively impact wildlife, habitat, and other natural resources, golf course managers should be encouraged to learn as much as they can about habitat, wildlife, and wildlife management practices. No one can cover an area as thoroughly as one who knows it, loves it, and lives in it day by day. To that end, what follows is a brief overview of wildlife biology and habitat management for golf course managers in the hope that their interest will take them further and provide them greater appreciation for and understanding of the wildlife and the land that they manage.

Wildlife Basics

The continuation of the conservation movement has become apparent in recent years through the increased pressure on every form of development and land use for greater attention to the environmental impact on wildlife habitat. Consequently, it has become increasingly important for golf course managers and other land managers to become more aware of the implications of their land management techniques. Because of public pressure, state and federal regulations, and the growth of the golf course industry, golf course managers have had to become more knowledgeable about how to manage wildlife on their golf courses. However, most golf course managers have not been trained as wildlife biologists. Because this book is intended to be a guide for golf course managers, it is important that you have at least a very basic knowledge of common groups or classes of wildlife and their characteristics.

The term "wildlife" is generally used to describe living creatures that are not domesticated, including mammals, birds, fish, reptiles, and amphibians. All of these species have a backbone and are known as vertebrates. Wildlife management is a means by which we protect, conserve, and sometimes control wildlife. If we have done our job well, wildlife management techniques are based on careful observation and data collected about various species. However, all wildlife management efforts are governed by a set of natural laws that were not created by humans. In order to better understand these natural laws and work with them to achieve desired wildlife management results, we must understand the basic types of wildlife species we want to manage.

In managing wildlife, one of the most important biological characteristics to understand is body temperature. It impacts the basic requirements for types and quantities of food sources and shelter to sustain the energy level required for any species to survive. All wildlife species are divided into warm-blooded and cold-blooded animals.

Warm-blooded refers to the ability to maintain a relatively constant body temperature irrespective of outside air and water temperature. For most mammals, being warm-blooded requires stores of food energy on a daily basis to stay alive. Human body temperature is normally 98°F. In order to survive cold, we have the ability to put more clothes on; or to survive the heat, to take clothes off. Some mammals may hibernate to lower their body temperature and reduce their need for food. Animals that remain active during winter in cold climates are on a tight energy budget and risk starvation if they cannot meet their energy needs and shelter requirements.

Cold-blooded means that body temperature fluctuates with the external temperature. Therefore, food energy is not spent on keeping warm, and the species can go a much longer time without eating. Likewise, activity levels vary with temperatures. During winter cold most cold-blooded animals are inactive or hibernating. Some species, for example reptiles, cannot survive freezing temperatures and must hibernate in holes or burrows below the frostline in climates with cold winters.

Since we are focusing primarily on wildlife that might inhabit golf courses, what follows is a broad discussion of birds, mammals, reptiles, and amphibians. We hope it will give you a foundation for learning more about the specific wildlife that inhabit your golf course and encourage you to seek out resources in your area that will help you expand your knowledge about the wildlife that surrounds you.

Birds

Humans have been fascinated by birds and their flight since the beginning of time. Icarus, in the Greek legend, fashioned wings so he could travel through the air like the birds. The Romans tried to foretell the future by "reading" the flight of birds. Columbus, near the end of his long voyage, correctly "read" the flight of birds winging west and followed them to the discovery of the New World. Today there are about 9,000 different species of birds in the world. About 680 species of birds are grouped into 73 families occurring annually in North America.

A bird is an animal with feathers. All birds have feathers; there is no animal with feathers that is not a bird. Feathers on the body provide insulation for the bird. Wing and tail feathers aid in flight. Most, but not all, birds fly. It is generally presumed, however, that those that do not are descended from ancestors that did. The bodies of birds are specifically designed for this amazing ability. The bones are light, but strong, and some bones contain air sacks connected with the lungs. Wings—marvelous adaptations of the vertebrate forelimb—are designed to propel the bird through the air. In flightless form, they balance the bird as it runs, like an ostrich, or propel it through the water, like a penguin. The tail is made entirely of feathers, unlike mammals, who have bones running down the middle of the body. The tail is long and often forked in strong flying or soaring species, but short in most species that fly little.

What looks like, and is generally called, the "leg" is actually a lengthened foot with toes at the end. What looks like the "knee" is really the ankle—the true knee usually being hidden by feathers high up under the bird's body. Most birds have four toes—three forward and one, a big toe, behind. However, some birds, like the woodpecker, have two toes forward and two behind. In many perching birds, the toes lock automatically around the perch when the body is lowered so the bird can eat or sleep without falling off. Birds have no teeth. Instead they have a bill that is adapted for the bird's preferred diet and used to pare, crush, or seize the flesh, seeds, or insects upon which a particular species exists. Food is sometimes temporarily stored in a crop, an enlargement of the esophagus near the stomach. The gizzard, or the main stomach, is a tough organ, and the process of digestion is aided by stones or gravel that the bird swallows for that purpose.

Though there are a variety of learned or mimicked behaviors, bird behavior is largely governed by instinct. Ornithologists have studied a variety of complex, unlearned action patterns, the most notable of which is migration. For example, young shorebirds leave their summer homes independently of their parents and follow a route for thousands of miles that they have never traveled before to reach their winter homes. For most birds, this migration is caused by the need for warmer temperatures or more plentiful food sources.

Birds are warm-blooded and need a constant supply of energy. Because they do not hibernate, birds may be critically stressed for energy during the winter months in cold climates. Small birds eat insects, fruits, or seeds. Large predatory birds, like hawks and owls, eat mice and other small rodents. Habitat needs include tall grasses, brush, a variety of trees for protection and breeding, and as food sources, and various-sized perches at all vegetation layers, from understory to canopy. Nest sites can include tall grass, brush or thickets, and standing dead trees, and can be supplemented by nest boxes. In all cases, to ensure breeding success, nesting sites should be in areas away from human activity.

Mammals

Mammals are warm-blooded like birds, but often hibernate in winter to reduce the need for energy to keep warm. They are air-breathing, milk-producing vertebrates with hair. With some mammals, the hair may be in the form of bristles. Even whales, at least at the embryo stage, have some bristles. Mammals give their young a period of parental care, and all—with the exception of the platypus which lays eggs—produce their young alive. Most have several different kinds of teeth. The normal blood temperature range for most active mammals is between 90 and 104°F. In hibernating individuals, however, the temperature is reduced, and may be only a few degrees above freezing.

Most mammals have four feet and a tail. In seals, whales, and manatees, the four feet are transformed into flippers, and in whales the hind limbs have totally disappeared except for two small bones deeply buried in the body near the base of the tail. In a few species the tail is greatly reduced or missing. Early mammals

A bobcat sighted at Wilderness Country Club, Naples, Florida.
Courtesy of Audubon International.

Mother bear and two cubs crossing the Nicklaus North Golf Course,
Whistler, British Columbia. Courtesy of Audubon International.

had five digits on each foot. This number was gradually reduced on many species until, like the horse, for example, only one digit bearing the hoof remained. Human beings have retained the original five digits.

Despite the prolonged internal development of mammals' newborn, they are often quite helpless after birth. They require parental care for some period of time

A white-tailed deer finds good habitat at Castle Pines Golf Club,
Castle Rock, Colorado. Courtesy of Audubon International.

in order to survive. This affords the parent an opportunity to teach the young, and
the species is thus enabled to break away from the pattern of pure instinct and
stereotyped responses characteristic of reptiles. With the increase of learned responses, a way opens to a greater variety in behavior, which means greater opportunity for the use and development of intelligence.

Small mammals usually eat insects, fruits, and seeds. Large mammals may be
either herbivores, eating plant materials like grass and fruit, or carnivores, eating
small mammals and other vertebrates like fish or amphibians. Cover and shelter
needs include caves, burrows, brush piles, hollow tree stumps, or nests in trees.

Reptiles

Of all the land vertebrates, reptiles have enjoyed the longest stay on earth.
During the Mesozoic era, the age of dinosaurs, they ruled the land and sea and
air—brontosaurus in the swamp, tyrannosaurs and pteranodons in the air, and
ichthyosaurus and plesiosaurus in the sea. They were the largest and most powerful creatures earth has ever seen. The group of reptiles known as dinosaurs has
long been extinct, but other groups still exist—namely turtles, crocodilians, liz-
ards, and snakes. Compared with their seemingly endless era of 130 million years,
the subsequent reign of the birds and mammals has been short-lived, and that of
humans could be compared to six minutes out of a 24-hour day.

A reptile is a cold-blooded vertebrate, with scales or plates instead of hair, that
lays eggs on land, or that bears living young on land or in the sea. The tough, dry

Pond slider moves slowly through John's Island Club,
Vero Beach, Florida. Courtesy of Audubon International.

skin of scales or plates functions primarily to preserve body moisture. Unlike amphibians, reptiles are not dependent on returning to water to breed. Because reptiles are cold-blooded and their temperature varies with the climate, they cannot survive freezing temperatures and are unable to survive in the arctic.

Reptiles exercise temperature control by seeking shelter from low or high temperatures. Hibernation is a means of avoiding the cold, and nocturnal activity a way of avoiding heat. As a result, they need brush piles and other natural low cover for protection. Reptiles usually eat insects or small mammals.

Reptiles are distributed all over the world, but they are more abundant both as species and as individuals in the tropics and subtropics than in the temperate zones. Some live on land, and others dwell in trees. Some inhabit fresh water, and a few enter saltwater.

Amphibians

The first amphibians arose in the Devonian period about 340 million years ago. For 140 million years, they were the dominant form of life on land. Some species grew to six feet or more in length, and their ancestral fish scales developed into protective armor. Today there are two orders of amphibians that occur in North America—salamanders, and frogs and toads.

The word amphibian is based on Greek words meaning "living a double life." Amphibians are cold-blooded vertebrates with limbs instead of fins, no claws on their toes, and moist skin instead of scales, hair, or feathers. Amphibians were the

Bullfrog habitat at the Village Links of Glen Ellyn, Glen Ellyn, Illinois.
Courtesy of Audubon International.

first vertebrates on land. Most of them still live part of their lives in the water. They are descended from fish, and most still have gills at some stage in their lives and must return to the water or to some damp place to lay their eggs. Amphibians breathe through their skin, gills, or lungs, or a combination of those.

Many amphibians spend the first part of their lives in the water as tadpoles. Later their bodies transform and they emerge to live the rest of their lives on land. Some spend almost their entire lives on land. Others, like mudpuppies, spend all their lives in the water. Amphibians do not drink, but absorb the moisture they need from the water or from damp earth. This permeability of skin makes them especially sensitive to toxic pollutants. The combination of aquatic and terrestrial life stages exposes them to the hazards of both. Therefore, they are good predictors of habitat quality.

Except in regions of extreme cold or dryness. Most are sensitive to temperature changes and many hibernate during cold weather. They tend to assume the temperatures of their immediate surroundings, so that during climatic changes they can endure wide ranges of heat and cold. Many avoid extreme heat or drought by burrowing into the ground. Amphibians are also sensitive to varying degrees of light and humidity.

Amphibians play an important part in keeping down the vast populations of injurious insects, such as mosquito larvae, because such insects form the chief item in their diet. They need standing bodies of water like marshes and swamps, and in woods, adults need covered objects like logs on the ground for protection and shelter.

Shallow ephemeral wetlands (those that are temporary—usually created by springtime runoff from higher elevations and rain) are more beneficial to amphibians, as well as other wildlife species, than deeper, permanent ponds. Amphibians also benefit from deeper ponds, where there are extensive shallow "shelves" and emergent aquatic vegetation.

Threatened and Endangered Species

Although excessive hunting and fishing once caused serious population declines and the extinction of some game species, today most important game animals are protected by strict regulations. The destruction of wildlife habitats is now the most important cause of declines in wildlife populations. Because many endangered species have very specific habitat requirements, they cannot adapt to new habitats if their original habitat is altered or destroyed. In addition to habitat destruction, environmental pollution, including pesticides, may in some cases contribute to declines in wildlife populations.

Endangered species are those plants and animals that are so rare that they are in danger of becoming extinct. When the population of a species drops to a certain low level, the remaining individuals may not be capable of reproducing sufficiently to ensure the survival of that species. **Threatened species** are plants and animals whose populations are very low or decreasing rapidly. They are not endangered yet, but are likely to become endangered in the future. As of 1999, 1,181 U.S. species are listed as threatened or endangered, of which 478 are animals. Plants and birds have the most listed species, followed by fish, mammals, and clams and mussels.

Believing that our rich natural heritage was of "aesthetic, ecological, educational, recreational, and scientific value to our Nation and its people," the Congress passed the 1973 Endangered Species Act to protect threatened and endangered plant and animal species. The purpose of the Endangered Species Act is to conserve "the ecosystems upon which endangered and threatened species depend" and to conserve and recover listed species. Under the law, species may be listed as either "endangered" or "threatened." All species of plants and animals, except pest insects, are eligible for listing as endangered or threatened.

The Endangered Species Act represented a bipartisan response to the decline of wildlife species and is regarded as one of the most comprehensive wildlife conservation laws in the world. The law is administered by the Interior Department's U.S. Fish and Wildlife Service and the

Commerce Department's National Marine Fisheries Service. Each year the United States Fish and Wildlife Services publishes a list of threatened and endangered species.

The study, appreciation, and conservation of wildlife is an ongoing process. In most cases people do not understand or appreciate the role that wildlife plays in the environment. Unfortunately, many species have become extinct before their role in the environment was adequately studied. That continues to be the case, as more species are added to the list of endangered and threatened species.

There is a delicate balance among animals and habitat and people. The more we learn about the various species that inhabit the land around us, the greater the chance that we will be able to maintain that balance, not only for the benefit of the wildlife itself, but also for our own benefit. The more we know about various species, the better we will be able to protect and enhance habitat that will prolong their existence and ours. Now that you have at least a cursory understanding of wildlife, a review of the fundamentals of habitat will move you a bit closer toward integrating your knowledge of wildlife and their requirements and creating and managing projects on your own property.

Wildlife Management Basics

We are blessed with a bountiful land supporting rich and diverse plant and animal communities. However, part of our natural heritage slips away each year. Development in general, agriculture, technology, transportation, and urbanization all affect wildlife, and its natural environment has become increasingly hostile. The land use decisions made by landowners and land managers are the key to the number of wildlife species and the number of individual animals found on their land. Decisions about the use or management of the land that are made for economic, aesthetic, or other reasons will always affect wildlife habitat and wildlife itself.

Although there are a variety of ways land can be used, from developed (residential, commercial, recreational, and educational) to undeveloped (preserves, refuges, "forever wild" forests, and national parks), there are natural laws, principles, and elements of land management that remain constant no matter how land is used or how it is managed. In order for you to manage the land that you oversee, you need to have a basic understanding of these concepts.

Wildlife is governed by natural laws, as distinguished from man-made laws. Natural laws are fixed and unchangeable. They are enforced by such basic factors as birth, death, and the need for food, water, shelter, and living space. Natural laws cannot be remade to suit the needs of humans. All that people can do is understand and work with these laws. To work against nature is to court certain failure.

One of the first basic truths in wildlife management is that animals are found almost everywhere. They are present naturally; they are not produced artificially or "made." Another fundamental truth is that wildlife, like all living creatures, must eat, drink, and find space and shelter to produce and raise young and protect themselves from predators. Wherever wildlife flourishes, one can be sure that those that are present have found the basic requirements for survival. Otherwise, they would not be there. When wildlife has the right combination of the basic requirements of life, they will flourish. On the other hand, when they do not find these conditions present, they will either starve for lack of food, die of thirst, or be eliminated by predators or natural elements for lack of cover. When wildlife cannot hide or raise young to replace their loss, they will have to move to a better environment or die. These are expressions of natural laws in operation that we need to understand fully in order to gain a better understanding of how to manage the land with a greater sensitivity to the needs of wildlife.

Basic Wildlife Requirements

Wildlife habitat is comprised of four basic components; that is, all species must have certain basic things and they must have some combination of them all through the year. They need **food** to eat. Food preferences are different for different species. Some eat plant materials, some eat seeds and grains, some eat insects, and some eat other animals. Some wildlife will eat a great variety of foods, while others eat only a few kinds. All species must have **water** to drink. Water sources are extremely important, and they are frequently the cornerstone for species survival. Water may be available from what humans see as common sources—ponds, streams, rivers, lakes, wetlands, marshes, and swamps. Other species are less dependent on common water sources because they get enough water from dew or from their food. Species also need **shelter or cover** to protect them from weather and from enemies so they can carry out the important activities of life—breeding, nesting, feeding, resting, and travel. They also need a certain amount of **living space**. This living space is an area in which wildlife will tolerate few if any of its own kind. This area can be a few square feet for a field mouse, or a few thousand acres for a bear.

Habitat at Breckenridge Golf Club, Breckenridge, Colorado.
Courtesy of Audubon International.

Requirements of each species will be different, but whatever the requirements are for certain species, all four must be met within a specific area or the species will not exist in that area. Many wildlife species have special ways of meeting their four basic needs. In order to avoid a winter scarcity of food, for example, some animals hibernate, others migrate to where there's enough food, and others store food.

Basic Concepts of Wildlife Management

In addition to the basic requirements of wildlife, it is important to understand some basic concepts relative to wildlife and wildlife management. First, we need to understand just exactly what habitat is. **Habitat** is an area that fills the four basic needs of any particular species. Some habitats are obvious. A stream is fish habitat and a hayfield is woodchuck habitat. Habitat requirements may be very specific—for example, fish habitat can be further divided into bass habitat and

Wetlands habitat at Old Marsh Golf Club, Palm Beach Gardens, Florida.
Courtesy of Audubon International.

Aspetuck Valley Country Club, Weston, Connecticut.
Courtesy of Audubon International.

trout habitat. Habitat is the most important influence on wildlife. When you become more familiar with wildlife and wildlife habitat, you will be able to locate a likely place to look for specific wildlife just by looking at the habitat. Sometimes certain conditions—such as temperature, humidity, and intensity of light—are important to a particular species and must be present in addition to the four basic needs in order to make an area suitable habitat for that species.

Wildlife habitat is not just trees, shrubs, grass, or even open fields. It is a complex mixture of plant communities or cover types. All play a role in meeting the needs of particular species, and all must be present within the species' normal range for that species to be present. The arrangement of cover types or plant communities, called **interspersion**, is also important to wildlife. While two equally-sized units of land can have exactly the same types of cover and be managed in the same manner, the wildlife populations they support may be very different, depending on how the cover types are arranged on that given piece of land.

In addition to the arrangement of plant communities within a particular piece of land, the connection between different habitats is also critical to a variety of wildlife species. Junctions between communities where a particular type of vegetation meets a different type of vegetation is called **edge**; for example, the border between woods and field, or the border between marsh and meadow is edge. Edge supports more wildlife than habitat that is exclusively woodland or meadow, or marsh. This is the **edge effect**—an increase in wildlife populations near an edge. Therefore, woodland edges should be allowed to develop low-growing trees and shrubs that will provide valuable habitat and increase both the number and types of wildlife species using the area. This is especially true for songbirds.

Wildflowers and edge habitat at Breckenridge Golf Club,
Breckenridge, Colorado. Courtesy of Audubon International.

Studies have shown that woodlands with well-developed shrub edges will support up to 95 percent more birds and up to 40 percent more bird species than those without edge development. Because of the vegetative and habitat diversity, edges are preferred nesting sites for both game and nongame mammals and birds.

A **patch** is any area of relatively uniform habitat, like a meadow or a cluster of trees. Small patches that are close together or connected with similar habitat may serve the same function as a single large patch. Large areas or patches are better for wildlife than small patches. A **corridor** is habitat of a particular type that differs from the adjacent land on both sides. It connects two patches of similar habitat and is used for traveling between patches.

Source-Sink

Source-sink is a term used in connection with bird conservation. It refers to the results of habitat management. While you may see birds on your golf course, the question is, Are they nesting, raising young, and therefore contributing to the overall health and vigor of the species?—that is, is your property a "source" or has your property only added to the overall decline of valuable habitat in the region, thereby becoming a "sink?"

Golf course architects and managers can have a positive impact on the wildlife value of their golf course by protecting or enhancing the

Maintaining patches and corridors at Pelican's Nest,
Bonita Springs, Florida. Courtesy of Audubon International.

property in ways that foster nesting activity and newly fledged birds
that will grow to maturity and raise their own young. For instance, many
species of birds depend on large blocks of forest interior woodlands.
These species are subject to predation by birds that are found more
commonly in open areas, and they have a low tolerance for human
activity around nesting sites.

By protecting and reducing maintenance in and around wooded
areas, allowing nature to take its course, and reducing human activity,
you can help to protect a variety of bird species. By protecting large
blocks of woodland habitat, you can play a critical role in ensuring
that the land becomes a "source" of vitality, rather than a "sink" that
drains away valuable habitat and wildlife.

Some species require large, relatively undisturbed blocks of specific types of
habitat, like older-growth forests or grasslands. The very existence of a species in
a specific area depends on habitat, but at any given time any habitat in an area has

a fixed limit for the kind and number of animals that may live in it; that is, any piece of land has a limited amount of space, cover, water, and food that will support a particular species. The number of animals that can be supported, or "carried," by the habitat is called the **carrying capacity**. For instance, a golf course has a specific carrying capacity because it is limited in the number of birds or mammals it can supply with food, space, cover and shelter, and water. However, by improving or enhancing the habitat, the number of animals it can support can be increased. The carrying capacity of habitat is usually lowest during late winter in cold regions; that is, the necessary requirements the land can provide for wildlife are diminished. Water sources may freeze over, available vegetation may decrease, or some cover and shelter may be lost. The amount and quality of food, cover or shelter, and water, which help to determine carrying capacity, are in turn determined mainly by: 1) **soil fertility**, and 2) **land use**.

Fertility is the richness of the soil. Fertility is determined by the kinds and amounts of food elements—like iron and calcium and nitrogen—that a particular soil possesses. These elements must be present in forms that can be used, first by plants and then by animals. If elements that help to grow bones and teeth or make blood are in short supply in the soil, animals that depend on such soils and resultant vegetation will be few and unhealthy and will fail to bear healthy young.

Land use that benefits people while conserving soil, water, minerals, and native plant and animal life is the best type of land use. Some land may be used for grain, while other land, which cannot be cultivated without erosion, may be best suited for pasture or meadow. Still other land, unsuitable either for cultivation or grazing, may grow valuable timber. Land use can be geared to wise practices so fertility can be built up, and the land conserved, and any land use can be enhanced for better balance with nature.

Besides the basic requirements for wildlife survival—food, cover or shelter, water, and space—and the basic concepts of interspersion, edge effect, habitat corridors, and carrying capacity, there are a number of other basic concepts that will help golf course managers better understand managing the land for wildlife.

Each species plays a slightly different role, or **niche**, in nature or in a particular environment. Even though a chipmunk and cottontail rabbit may live on the same plot of ground, they don't eat the same foods, and therefore they occupy different niches. Any difference in food, habitat use, nesting behavior, or other features means the niche is different. Niches of two species may be similar, but they are hardly ever the same.

Many animals maintain a **territory** at certain times. This is an area usually right around the home or nest that the animal will defend against members of its own species and occasionally against other species. Many birds share the tree in which they nest with birds of other species, but chase off any of their own kind. If a male pheasant defends a territory of three acres, and you have a six-acre field, you will have only two male pheasants—no matter how much food, water, and shelter are available. In this case, living space is a limiting factor.

A **limiting factor** is the basic requirement—food, water, cover or shelter, or space—that is in shortest supply, and consequently prevents the wildlife population from getting larger. If a golf course has a bird feeder with seed for food, ponds for water, and plenty of open space, but no shrubs, hedges, or trees nearby, what is the limiting factor for birds? It is shelter or cover since there are few places that will help protect birds from predators or the weather and no place for nesting. In this case, the population will be kept at a low level because there will be increased mortality from predators and bad weather. Knowing what wildlife exists on your property, understanding their wildlife requirements, and assessing limiting factors are essential elements in making wildlife management decisions.

Golf course managers can help wildlife by understanding and working with nature and its laws. They can increase wildlife by following land management practices that will improve the carrying capacity by providing more and better food, water, and shelter for protection or breeding and nesting, and enhancing the vegetative diversity of the land.

Concerned golf course managers can reverse the trend of decreasing habitat and wildlife by taking action on their property. Depending on the type and condition of habitat on the property and the wildlife species you would like to attract or increase, your plan might be as simple as constructing brush piles, or as complex as designing a grass-shrub-tree planting several acres in size. All habitat improvement projects can be reasonably approached through four basic steps:

1. Determine the species of wildlife that live in the area.
2. Identify the species you want to manage for and learn their habitat requirements.
3. Identify the habitat elements lacking or in need of improvement.
4. Design and implement projects that establish or improve wildlife habitat.

Approaches to Wildlife Management

This book is intended to address the relationship between golf course and wildlife management by providing a historical, ecological, and practical foundation. It is important for golf course managers and others in the golf course industry to be familiar with conservation and land management activities in a broader perspective. Regardless of the type or setting of wildlife habitat, there are a variety of approaches to habitat management that can enhance the conservation of wildlife as well as provide enjoyment and education for people. Some of these approaches are also applicable to golf courses. Historically, the following approaches to wildlife management have been taken singly or in combination:

1. Preservation, enhancement, or creation of natural habitat.
2. Legislation and enforcement to:

Levels of vegetation.
Courtesy of Audubon International.

    a. provide protection for, or
    b. regulate the harvest of certain wildlife species.
3. Establish refuges or sanctuaries to:
    a. protect wildlife species,
    b. provide wildlife with places to reproduce, or
    c. provide educational opportunities to observe wildlife.
4. Control predators to reduce predation on:
    a. certain wildlife species, or
    b. livestock.
5. Artificial propagation to stock selected areas, such as:
    a. hatcheries for certain species of fish, or
    b. game farms for certain species of wildlife.
6. Provide and enhance food sources for birds and other wildlife, especially during winter.
7. Establish man-made structures to supplement habitat, such as:
    a. erecting nest boxes for nesting species, or
    b. providing brush and rock piles for additional cover.

Integrating transition areas for golfers at Pumpkin Ridge Golf Club, Cornelius, Oregon. Courtesy of Audubon International.

8. Habitat management in urban and suburban areas, including:
   a. vegetation management (including eradicating invasive vegetation) and water manipulation, or
   b. special architectural designs, placement, or construction of buildings.

Because these are broad approaches, golf course managers will have to decide which are appropriate to their location and situation. From an environmental and wildlife management point of view, particularly as it relates to golf courses, there are some additional suggestions that may help in determining what you can do on your course.

Your wildlife management goal should be to have as many "blocks" of natural habitat as possible. These blocks, no matter what size they are, should be connected by wildlife habitat corridors. These blocks of habitat and the corridors should be comprised of plants that are native to the region in which the golf course is located. In addition to using native vegetation, it is important that habitats have horizontal and vertical structure. If your property is in a forested area, for example, create, enhance, or *preserve vegetation in various heights* from understory to shrubs, to smaller trees, to tall mature trees.

The key to managing golf and wildlife habitat is defining areas on the golf course that may prove to be conflict sites between the game of golf and your wildlife habitat conservation efforts. Think in terms of zones or corridors, and transition areas for golfers, just as we have discussed for wildlife. For golfers, there needs to be ample and wide landing areas on the fairways and transition areas, and then out-of-play areas. Tees and greens are intensively maintained with frequent human activity. These are not good wildlife habitat areas. We also know that the majority of golfers play right-handed, and as a result tend to hit a golf ball from left to right. Therefore, putting thick tangles of good wildlife habitat on the right-hand side near a fairway landing zone is probably not a good idea. But on the typical 150-acre golf course, there are many opportunities to judiciously incorporate and expand wildlife habitat, and manage vegetation to maximize food, cover or shelter, and space for wildlife in out-of-play areas. Projects that enhance habitat need not detract from the game of golf, but can positively impact both the wildlife use of the course and the aesthetic nature of the game.

Landscaping for Wildlife

As a golf course manager, you can play an important part in providing essential elements that positively impact a variety of wildlife, including birds, amphibians, reptiles, and mammals. By identifying parts of your golf course that could become more natural and less highly maintained and by carefully choosing trees, plants, shrubs, and flowers for areas around your course, you can effectively provide habitat that satisfies wildlife's basic needs for survival.

Naturalizing your landscape, or landscaping for wildlife, means understanding and making decisions about the vegetation around your course. What plants grow on your course? Where do they grow? How intensively do you have to maintain them? Most importantly, what function do they serve?

We know that the primary portion of the golf course serves as a playing surface for the game of golf, and we know that turfgrass for tees, greens, fairways, and roughs grows there. We also know that it is intensively maintained and that this is dictated by its function. But what about the other parts of the property that are not turfgrass? What trees, shrubs, flowers, long grasses, wildflowers, or "weeds" grow in those areas? What function do they serve, and how do you maintain or manage them? Is there a way you can reduce maintenance and provide some value for wildlife at the same time?

Absolutely! There are many opportunities for you to incorporate more natural landscaping around your golf course. You can include plants and flowers around the clubhouse and the maintenance facility, along the entrance roads to the golf facility, and around parking lots. More importantly, you can identify areas around the golf course where the game is not usually played, and target them as areas where you can reduce maintenance. You can choose grasses, plants, and trees that will not only add to the visual beauty of your landscape, but provide wildlife habitat for birds, butterflies, mammals, and beneficial insects that play a critical role in the ecology of your golf course.

Natural landscaping also plays an economic role. Today's golf course maintenance is "high tech" and it costs money to keep up with the latest technology. Many golf courses have invested in expensive mowing equipment, weather stations, and computerized irrigation systems. Water, electricity, chemicals, equipment maintenance, gasoline, and labor are only some of the costs attributable to turfgrass maintenance, and it makes sense that while turfgrass is an integral part of the game of golf it should be used only in areas where it needs to be used.

So in areas where you don't absolutely have to have turfgrass, reduce your costs by reducing your maintenance practices. By using the right native vegetation in appropriate locations, you can create blocks of habitat and habitat corridors. Identify areas that are currently mowed on a frequent basis, or that are difficult to maintain, like severe slopes, around stands of trees, or around water bodies that are not part of where the game is played, and replace those areas with native vegetation. Allow them to become more naturalized and you will not only enhance those areas for the wildlife on your course, but you will also save money in the process.

This chapter is designed to give you some ideas and basic information about landscaping for wildlife that have a fairly wide application. There are a variety of suggestions for the types of plants you can use that provide food and cover sources for wildlife, as well as some suggestions for structures you can create that will also enhance wildlife habitat. But this will only serve as a guide for you. It is not possible to detail all of the various plant types, habitat types, soil types, and local or regional weather conditions that impact landscaping for wildlife. There are, however, a variety of publications and resources available at bookstores and libraries that will provide you with specific local information.

You will have to do some of your own research. You will have to determine the exact types of plants, habitats, and wildlife that are found in your region and decide what will grow best where. You will have to determine what will work for you in planning how to enhance the habitat on your course and still meet the expectations of the golfers. We know it is a delicate balance sometimes, but we also know many golf course managers have successfully naturalized their golf courses with the growing support of golfers. It is the "investigative" part of the process that establishes the foundation for an ultimately successful management program. You should know more about your site than anyone else. It will be up to you to teach others what you have learned, what you would like to do, what you have done, and why it is important for the game and the environment.

We have discussed the four basic requirements that are applicable to all wildlife—food, water, shelter or cover, and space, and in the chapter "Wildlife Management Projects," we will discuss some wildlife habitat management projects that you can implement on your golf course. All of the projects are founded on two basic types of landscaping components, and these components help fulfill the four major habitat requirements of wildlife. The discussion of **plant components** and **structural components** of wildlife landscaping will provide you with an overview of the interrelationship between wildlife and the natural landscape.

Water cooler surrounded by naturalized plantings.
Sonnenalp Golf Club, Edwards, Colorado.
Courtesy of Audubon International.

Plant Components of Wildlife Habitat

Plant components should be organized and managed by the season during which they provide major food and shelter value. The golf course landscape that includes plants from the following categories will provide year-round benefits to a wide variety of wildlife. The appropriate vegetation to use in your area and the exact species of wildlife that may be found in your area can only be determined by a measure of research and planning on your part. The plant names we have suggested are common in many areas of the United States, but may not be suitable for your area. Appropriate plants for your ecological region may be found in the *Landscape Restoration Handbook,* or other gardening or planting references, or suggestions may be made by your local nursery.

Remember that diversity is the key to valuable wildlife habitat. People often tend to simplify the landscape by creating large areas, or monocultures, which contain a single species. These are quite vulnerable to environmental problems. When you increase the number of plant species used in a landscape plan, you increase the ecological stability of the golf course. In addition, many plant species support an abundance of wildlife. Such habitats are less vulnerable to large-scale destruction caused by insect pests or disease, which can devastate a single plant species. So in terms of plant components, diversity is essential.

The American Chestnut

The American chestnut was originally a large tree, but now exists primarily as sprouts from old stumps. At one time, it was a dominant tree in dry forests throughout much of the United States. After 1900, however, a fungus bark disease, believed to be of Asiatic origin, became epidemic and completely eliminated the American chestnut as an important forest tree.

Chestnut lumber was quite valuable. It was used for furniture, musical instruments, interiors, caskets, and fences. Tannin was derived from the bark, and the nuts were marketed. Bobwhite, wild turkey, squirrels, and white-tailed deer are among the many species of wildlife that once fed on the nuts.

Sprouts may continue from some old stumps, and these flower and produce fruits. As soon as these shoots attain a moderate size, however, the disease attacks the bark at the bases of the trees. It is hoped that someday a blight-resistant specimen may occur, from which a new strain could be developed.

Reference: *A Field Guide to Trees & Shrubs*, George A. Petrides, The Peterson Field Guide Series, Houghton Mifflin Co., 1972

Note: The delineation of plant and structural components and suggested plantings was drawn from the publication, *Landscaping for Wildlife*, produced by the Minnesota Department of Natural Resources.

Plant components of wildlife habitat can be divided into six areas:

1. Conifers
2. Grasses and Legumes
3. Butterfly, Bee, and Moth Plants
4. Hummingbird and Oriole Plants
5. Seasonal Plants: Summer, Fall, and Winter
6. Nut and Acorn Trees

1. Conifers

Conifers, also known as evergreens, provide protective winter shelter, summer nesting cover, and food for a variety of species. They include trees and shrubs that generally do not lose their needles in the winter and so provide protective cover. The branches and cavities are frequently used as nest sites. Pines, spruces, firs, junipers, cedars, and yews are conifers that provide exceptional food and cover. The sap, needles, twigs, buds, and seeds are eaten by wildlife. The eastern white pine, for example, is used by 48 species of birds. Yellow-bellied sapsuckers eat the sap, spruce grouse and turkeys eat the needles, and many other birds eat the seeds. The Canada yew is so highly preferred by deer that it is generally wiped out when significant numbers of deer are present. In addition to providing food and cover for wildlife, conifers provide visual enjoyment by staying green all year round.

Tips for Planting Trees on Golf Courses

- Locate trees so they do not conflict with the water or sun requirements of turfgrass.
- Anticipate growing tree roots when planting near paved surfaces such as cart paths and parking lots.
- Choose a variety of trees that are native and appropriate to your region rather than nonnative trees.
- Plant trees in groups rather than as single trees to increase their wildlife habitat value.

---

2. Grasses and Legumes

Grasses and legumes provide nesting and winter cover and some food sources for a variety of species, including ring-necked pheasants, mallards, blue-wing teal, meadowlarks, dickcissels, bobolinks, and vesper sparrows. Grasses and legumes also provide forage for plant-eating animals, such as white-tailed deer, cottontail rabbits, woodchucks, meadow voles and others.

Besides providing critical transition areas between forest and open field, grassy areas provide summer cover for ground nesting birds accompanied by young, as well as hunting sites for red fox, red-tailed hawk, kestrel, northern harrier, short-eared owls, coyotes, and long-tailed weasels. In addition to providing spring and summer nesting cover, switch grass provides winter cover for pheasant and deer, and food for seed-eating winter birds like the American goldfinch.

Legumes play an important role when planted with other grasses because they convert or "fix" atmospheric nitrogen in their roots. This nitrogen subsequently becomes available to other plants as a source of natural fertilizer. As with other plantings, a solid stand of switchgrass will have less value than creating a more diverse habitat. Recreating a prairie habitat by planting a variety of native grasses, legumes, and flowers will not only enhance the value of the habitat, but will add to its visual appeal.

Naturalized area at Sonnenalp Golf Club, Edwards, Colorado.
Courtesy of Audubon International.

3. Nectar Plants for Bees, Moths, and Butterflies

Planting nectar plants that attract butterflies, bees, and moths provides a wonderful opportunity to not only enhance your golf course for wildlife, but also to provide aesthetically appealing gardens around your clubhouse. You may wish to develop gardens using native prairie wildflowers or native woodland wildflowers to attract butterflies like monarch, painted lady, comma, red spotted purple, tiger swallowtail, several fritillaries, red admiral, sulfurs, cabbage butterflies, and several species of blues. In addition, your plants can provide nectar sources for honeybees and bumblebees, hummingbirds, clear-winged moths, and sphinx moths.

Butterfly Plantings: Two types of food are necessary for butterflies—food for caterpillars and nectar sources for adult butterflies. Like other insects, butterflies are cold-blooded. They rely on the sun to keep them warm and raise their metabolism so they can fly. In addition to plant sources for food, you can also provide stones for sunning and a shallow water source.

Monarch butterflies on coneflowers at Salina Country Club,
Salina, Kansas. Courtesy of Audubon International.

Butterfly Plantings

Trees	birches, aspens, willows, hackberry, cherries, and oaks
Plants	composites, legumes, grasses, herbs, blueberries, sedges, and docks, including asters, alfalfa, vetches, clovers, Kentucky bluegrass, little bluestem, and violets
Larval food sources	hollyhock, milkweed, lupines, black-eyed Susan, sedum, and marigolds
Nectar source plants	dogbanes and milkweeds, asters, thistles, goldenrods, Joe-Pye weed, ironweed, fleabane, red clover, wintercress, selfheal, vetches, peppermint, globe thistle, purple coneflower, and blazing star

Bee Plantings: There are a variety of plants that are excellent for attracting bees. Among the most significant bee plants are those that are available when bees first emerge in the spring. Many of the best bee and butterfly plants are herbs. Creating an herb garden not only benefits wildlife, but the human palate as well.

Bee Plantings	
Spring plants	grape hyacinth, jonquil, daffodil, sweet mock orange, cherry, apple, plum, peach, apricot, almond, pussy willow, and lilac
Other plants	evening primrose, penstemons, petunia, phlox, moss rose, salvia, sedum, goldenrod, globe thistle, obedient plant, coralberry, wolfberry, snowberry, marigolds, clovers, garden verbena, broccoli, and Mexican sunflower
Herbs	borage, hyssop, lavender, mint, spearmint, peppermint, applemint, lemon balm, sweet marjoram, wild marjoram, rosemary, sage, dill, winter savory, and thyme

Moth Plantings: Several kinds of moths can be attracted regularly to a flower garden.

Moth Plantings	
Hummingbird clear-winged moth (resembles bees)	attracted by sweet William, fireweed, dame's rocket, bergamot, showy evening primrose, petunias, sweet mock orange, and phlox; coralberry and snowberry provide larval food
Night-flying sphinx moth (resembles hummingbirds in flight)	attracted by night flowering plants, including sweet William, heliotrope, dame's rocket, Madonna lily, white lilies, marvel of Peru, flowering tobacco, and petunia
Day-flying sphinx moths	attracted by trumpet creeper, dwarf blue gentian, standing cypress, Madonna lily, white lilies, cardinal flower, phlox, and old-fashioned weigela

4. Nectar Plants for Hummingbirds and Orioles

Nectar-producing plants are also attractive to hummingbirds and orioles. Hummingbirds are the smallest birds on earth. Because they move quickly and have a high metabolism, a hummingbird may eat more than one-half its weight in food and eight times its weight in fluids daily. They feed fourteen to eighteen times per hour for less than one minute and rest between meals. Flower nectar and tiny insects are their preferred diet. A good strategy is to provide some plants that bloom in early summer and some that bloom in late summer. Orioles may also be attracted to feed on nectar or blossoms of several red or orange flowers.

Plants attracting hummingbirds	early-blooming plants: American columbine, petunia, and foxglove; later-blooming plants: red (scarlet) bergamot, cardinal flower, and dwarf blue gentian
Plants attracting orioles	hollyhock, trumpet vine, daylily, lemon daylily, tiger lily, turk's cap lily, and scarlet trumpet honeysuckle

5. Seasonal Plants

Trees, shrubs, and vines that grow throughout the summer, fall, and winter provide a variety of food, nesting shelter, and protective cover for a diversity of wildlife.

Summer Fruit, Berry, and Cover Plants

From June through August, many plants provide food and nesting cover, and especially important are those that produce fruits and berries. These fruit and berry plants attract a wide variety of birds, including the brown thrasher, bluejay, gray catbird, American robin, eastern bluebird, wood thrush, cedar waxwing, Baltimore oriole, scarlet tanager, northern cardinal, indigo bunting, Eastern towhee, dark-eyed junco, various woodpeckers, grouse, and pheasant. They are also attractive to deer, squirrels, raccoons, red fox, and many butterflies.

Several plants offer an additional benefit for wildlife. They spread by growing suckers that create dense thickets, making ideal nesting cover for shrub-nesting species like catbirds and brown thrashers. These thickets can also serve as important winter pheasant cover. Thicket-forming plants include wild plum, scarlet Mongolian cherry, choke cherry, black raspberry, red raspberry, and blackberry.

Grapes and other vines can be used to enhance the value of fences or dead trees. If you plant vines at the base of a fence or dead tree, the vines can climb the fence or tree and subsequently create nesting cover and summer fruits.

Summer Plants	
Trees	red mulberry, choke cherry, and black cherry
Tall shrubs (15'–25')	Manchurian bush apricot, choke cherry, bird cherry, and serviceberry
Medium shrubs (10'–15')	American plum, Siberian plum, pin cherry, and Nanking cherry
Low shrubs (1'–10')	scarlet Mongolian cherry, sand cherry, lilac-flowered honeysuckle, raspberry, elderberry, blackberry, Juneberry and blueberry
Vines	grape (beta and riverbank)

Fall Fruits, Grains, and Cover Plants

Several shrubs, vines, and grain crops, including wheat and corn, are valuable sources of food in the fall, and sometimes into the winter if they are not immediately consumed by wildlife or covered by snow. Fall wildlife food sources allow migratory birds to build up fat reserves before they migrate, as well as benefiting resident nonmigratory wildlife species to help them survive the winter.

A variety of birds benefit from the fruits of the red dogwood, grey dogwood, mountain-ash, winterberry, cotoneasters, and buffaloberry. These include gray catbirds, brown thrashers, American robins, wood thrushes, cedar waxwings, cardinals, purple finches, dark-eyed juncos, black-capped chickadees, white-breasted nuthatches, evening grosbeaks, ruffed grouse, bluebirds, wood ducks, pheasants, and orioles.

Winter Fruits and Cover Plants

Many of the best winter wildlife foods are characterized by two important qualities: persistence and low appeal to wildlife when they first mature. Examples of persistent foods are glossy black chokeberry, Siberian and red splendor crabapple, common snowberry, staghorn and smooth sumac, bittersweet, American highbush cranberry, Eastern and European wahoo, and Virginia creeper.

Many fruits are not immediately desirable as wildlife foods. Some are bitter when they first ripen. Others must freeze and thaw several times until the fruit breaks down to become more palatable. Plants like bittersweet may not be eaten until late winter or spring. Plants such as sumacs and highbush cranberries are extremely important for late winter survival when other food supplies are limited or exhausted. If space is severely limited on your golf course, winter foods are probably the most important category of plants you can provide because natural foods are most limited during the winter season.

6. Nut and Acorn Plants

Nut- and acorn-producing plants are another extremely important landscape component for wildlife. Nuts and acorns, referred to as "mast," are significant food sources in the fall and winter for white-tailed deer, wild turkey, wood duck, pheasant, gray squirrel, fox squirrel, red squirrel, ruffed grouse, bobwhite, mallard, black bear, raccoon, and many other species. Examples of some important mast producers are white oak, burr oak, Northern red oak, live oak, water oak, American hazelnut, walnuts, shagbark hickory, and butternut.

Perhaps one of the greatest benefits of planting these hardwoods is that they are a wonderful long-term investment in wildlife production. Many of the fruit-producing shrubs are relatively short-lived, whereas oaks may produce acorns for up to 400 years. These trees also contain natural cavities that are used by up to a hundred species of wildlife.

Structural Components of Wildlife Habitat

The structural components of habitat can be built, placed, and created to benefit species that will use these types of structures. Some structures, like feeders, not only play an important role for wildlife, but can also be used to educate golfers about wildlife found on the golf course. Unlike plant components, which may vary depending on your geographical location, the structural components are the same and are applicable to the species of wildlife for which they were intended regardless of geographical location. There are six structural components of wildlife habitat.

1. Nest Boxes
2. Dead Trees (snags), Fallen Trees, and Perches
3. Brush and Rock Piles
4. Cut Banks, Cliffs, and Caves
5. Water Sources
6. Bird Feeders

1. Nest Boxes

Nest boxes and nest platforms are used by over 75 species of wildlife in the United States. On most golf courses, woodlot management of trees to preserve

Nest boxes have been placed around each wildflower area at
Stillwater Country Club, Stillwater, Minnesota.

Mounting nest boxes helps to balance the loss of
natural habitat for cavity nesting birds.

natural cavities is the best way to provide cavity nest sites. However, nest boxes are a supplement to the natural cavities found in dead trees. More detailed information about next boxes is included in "Wildlife Habitat Management Projects."

2. Dead Trees (Snags, Fallen Trees, and Perches)

A **snag** is a dead or dying tree that is left standing. To many people, a snag is just firewood waiting to be cut. Once a tree dies, the process of decay begins. As the hardwood softens, woodpeckers excavate nesting and roosting cavities. After they abandon these cavities, other wildlife may use them. Until recently, foresters systematically removed dead trees because of their potential for harboring disease and insect pests. This trend is changing. It is now widely recognized that many

Tree snags can provide cavities for nesting and perches.
Courtesy of Audubon International.

bird species that nest in snags are those that feed heavily on insects, thereby helping to prevent serious insect outbreaks. Most dead trees do not harbor active diseases or damaging insects.

Cavities occur in both living and dead limbs. Large cavities can occur where major limbs die and fall from the trunk. These damaged areas are usually insect-infested, and the cavities are enlarged by birds and mammals digging into the cavities and picking apart the wood to get at these insects.

Perches are another important habitat component. Kingfishers and herons use branches overhanging water as perches from which they can spot fish to feed on. Snags also serve as perches from which flycatchers fly out to catch insects, often over fairways. Snags or tree branch perches can be beneficially used over ponds, on the edge of fairways, or at bird feeders to stimulate additional bird use, so long as these snags do not pose safety problems for humans.

Den trees have trunks or large limbs hollowed out by rotting, with an opening to the outside. This includes some snags, of course, but den trees typically are still alive enough to continue to produce mast (nuts and acorns) or fruit. Den trees are used by honey bees, birds, and mammals varying in size from a mouse to a black bear. Hollow trees broken off at the top and open to rain and snow provide less protection, but are sometimes used by birds like great horned owls for nesting protection.

3. Brush Piles and Rock Piles

Brush piles and rock piles provide escape, cover, nesting sites, and den sites for rabbits, weasels, woodchucks, skunks, Northern prairie skinks, red foxes, garter snakes, and many other species. Brush piles can also provide important reptile, amphibian, and fish habitat if placed on the edge of a small pond so that part of the brush is submerged.

Brush piles should be placed in sheltered areas along the edges of fields, fairways, and woods. A brush pile should have a foundation of big rocks, stumps, and logs to keep it from decomposing too quickly. Several heavy logs can be placed on top of the pile to keep it from blowing apart. Old sections of culverts or sections of hollow logs placed within the foundation of the brush pile can serve as animal den sites.

Similarly, rock piles can be placed on the north side of ponds larger than one acre. Such rocks can be football-sized, up to 3 feet in diameter. They can be dumped along the water's edge, up the bank 3 to 4 feet, and below water level to the depth of 2 to 3 feet. These sites provide both aquatic shelter for frogs and toads and sunny basking sites for turtles, skinks, and snakes. Rock piles are excellent duck and turtle basking sites if they are placed away from the shoreline to make them less vulnerable to ambush by predators.

Snag and brush pile provide cover and nesting sites.

4. Banks, Cliffs, and Caves

While banks, cliffs, and caves may not be normal types of habitat associated with golf courses, if they exist as part of your golf course site, special attention should be given to their enhancement and protection. They are important features of the landscape for several species, ranging from peregrine falcons to kingfishers, bats, and bank swallows. Exposed soil, gravel, and even limestone banks along creeks and rivers are used as nesting sites by rough-winged swallows, belted kingfishers, and bank swallows.

Exposed banks and gravel pit areas are also used as den sites for badgers, red foxes, coyotes, and woodchucks. Preservation of these features is actually a good habitat management technique as long as they do not pose a safety hazard to humans.

Caves are a habitat feature important primarily for bats and a variety of specialized invertebrates. Should there be caves on the property, care should be taken to preserve them. They may need to be protected or sealed with a gate to prevent entry by people who might inadvertently disturb hibernating bats.

A small bird bathing pond and waterfall was created at the
Country Club of Florida, Village of Golf, Florida.
Courtesy of Audubon International.

Diverse habitat with buffered water source at River Run Golf Course,
Berlin, Maryland. Courtesy of Audubon International.

5. Water

Water is an essential component of wildlife habitat. It will attract a wonderful variety of wildlife, from songbirds to small mammals, reptiles, amphibians, and beneficial insects. Providing water is an integral part of landscaping for wildlife. Not all wildlife need standing or free-running water to sustain themselves, but even those that do not require these conditions still seem to prefer them. Dripping or flowing water seems to be more attractive to many wildlife species than still water. No matter how abundant food is on your property, without water for drinking and bathing many species of wildlife will not use your property.

Water can be provided in many forms, from a dripping source of water in a mud puddle for butterflies, to birdbaths, man-made ponds, natural springs, creeks, marshes, lakes, and rivers. Water sources such as springs, beaver ponds, marshes, creeks, swamps, lakes, and rivers are vital components of our environment. One of the biggest challenges in developing a landscape for wildlife is to preserve and manage water habitats where they still exist, to create ponds where they are absent, and to restore wetlands where they have previously been destroyed. One of the best ways of improving habitat is to enhance and protect wetlands, ponds, marshes, and other water sources. Wetland planting and aquascaping have the added benefits of stabilizing shorelines, reducing erosion, increasing aesthetics, and improving water quality.

The Value of Wetlands

For many years, wetlands were seen merely as "swamps" and "mosquito breeding grounds." Because their value to both people and wildlife went unrecognized, many wetlands became dumping grounds for garbage and toxic waste, or were drained and filled to make way for development projects.

We now know that wetlands play an integral role in providing wildlife habitat and ensuring water quality for essential groundwater supplies and downstream water resources. Here are just a few of the many benefits of wetlands:

- Wetlands provide breeding, nesting, and feeding grounds for a great diversity of wildlife, including fish, reptiles, amphibians, mammals, and birds. Shorebirds, waterfowl, and migrating ducks and geese are especially dependent on wetlands across North America, as are rare species, such as the bald eagle and osprey.
- Wetlands help prevent flooding by absorbing and storing stormwater. They also protect subsurface water resources and help to recharge groundwater supplies.

Wetlands play an important role for both humans and wildlife.
Courtesy of Audubon International.

- Wetlands are essential nursery grounds and sanctuaries for fresh-water fish. Aquatic plants and animals at all levels of the food chain thrive on the rich source of nutrients that wetlands provide.
- Wetland areas help protect and improve downstream water quality by absorbing silt and organic matter, and filtering pollutants from streams.
- Wetlands are vital gathering areas for many migratory birds.

6. Bird Feeders

The finishing touch to a wildlife landscaping plan is an assortment of bird feeders, which supplement the foods already provided by trees, shrubs, and flowers. Wildlife feeders can serve several functions. The enjoyment of observing birds by providing feeders in backyards has been a popular activity with homeowners for years, and has become increasingly popular with golf courses, where feeders are placed near the clubhouse. Although for the most part bird

Bird feeders provide additional food sources for birds during winter months. Courtesy of Audubon International.

feeding is not necessary to birds' survival, it is in some cases an opportunity to help birds get through a cold, icy winter day. In addition, it provides educational and enjoyable opportunities for people to see and learn about the various birds that are utilizing their property.

Enhancing Habitat Diversity on Your Golf Course

The plant and structural components comprise the foundation for enhancing your golf course property for wildlife. Rather than thinking about these components separately, think of them as parts of a larger concept. Both plant and structural components help supply additional sources of three of the four basic wildlife requirements—food, cover and shelter, and water—as well as increasing the habitat diversity on golf courses.

Creating more diverse habitat can be accomplished by "naturalizing" more of the golf course. Naturalizing the landscape means using native plants to restore and enhance the landscape and reducing the amount of human maintenance. It is not merely letting nature take over, but rather making reasonable decisions about how to enhance and manage the land from a more natural perspective. Remember, though, at first some golfers might see "naturalized" areas as neglected and unattractive, so it is critical that you communicate with them about what you are trying to accomplish and help them understand the reasons for and benefits of a more natural landscape.

One persuasive argument for a more natural landscape is the economic bottom line. Of course, there are initial costs associated with planning and establishing naturalized areas that are beneficial to wildlife. But, over time you will save water, reduce maintenance (less mowing, less gas, less labor, less wear on machines), reduce pesticide use, and enhance the visual appeal of the golf course.

Naturalizing Your Golf Course

Habitat diversity can be increased on nearly every golf course by naturalizing nonplay areas. Enhancing food and cover sources for wildlife will help you attract and sustain the greatest diversity of wildlife species. Techniques such as letting taller grasses grow, restoring woodland understory, and choosing native plants when landscaping not only increase the habitat value of your course, they also increase the overall acreage and diversity of wildlife habitat in general.

75

A planned naturalized area at Hominy Hill
Golf Course, Colts Neck, New Jersey.
Courtesy of Audubon International.

Begin by looking at your course from a wildlife perspective, assessing the basic requirements for survival—food, cover, water, and space. Parts of your property may already provide some of these elements. By naturalizing areas you can extend, connect, and build upon existing habitats.

Next, consider potential locations and types of projects you want to pursue. If your golf course is currently intensely maintained, you may want to start slowly, learn by your experiences, and gain golfer approval before you undertake large-scale enhancement or restoration projects.

- Location is the most important consideration in terms of plant selection, visual appeal, and acceptance by both the players and surrounding property owners. Look for nonplay areas between fairways, under small stands of trees, and along wooded edges. Begin slowly and increase the naturalized area.
- When making plant selections for trees, shrubs, and flowers, choose native plants that provide food and cover. Locating plants near water sources will increase their habitat potential. Because native plants are well adapted to your local climate and soil, they will require less maintenance. Survey your course and learn more about the native plant communities in your area to determine which species will grow best on your site.
- While native species are extremely tough and hardy, they do benefit and establish faster with some site preparation and postplanting care. Mulch-

ing, weed barriers, and supplemental irrigation will increase shrub survival rates.

We recommend using native plants when naturalizing golf courses. There are important differences between **native**, **naturalized**, and **exotic** or **invasive** plants.

Native plants originated and grow naturally in a particular region or habitat. Native plants are adapted to the climate and soils of an area and require a minimum of maintenance once they are established. They also provide good sources of food and cover for wildlife. You can establish native plants in flower gardens, habitat areas, around buildings, entry roads, or property borders.

Naturalized plants are established in areas distant from their origin. Many of North America's nonnative, naturalized plants, such as Queen Anne's lace, day lilies, Scotch pine, ox-eye daisy, and chicory, came from other countries and have established themselves successfully in this country.

Exotic or **invasive plants** are those that grow outside of their place of origin and are interfering with the growth of native species in that location. If you have exotic invasive plants on your property, such as kudzu, purple loosestrife, Japanese honeysuckle, and melaleuca, plan to cut them back or eliminate them as part of your habitat management plan.

A hand-planted constructed wetland with nest box at the
Carolina National Golf Club, Supply, North Carolina.
Courtesy of Audubon International.

Wildflower area incorporated as a buffer zone around a
pond at Aurora Country Club, Aurora, Illinois.
Courtesy of Audubon International.

If you need assistance incorporating native plants into your landscaping, you can consult landscaping companies, local horticulturists, native plant societies, the cooperative extension, or the Department of Natural Resources in your state. Check with nursery sources for native plants. Not all nurseries sell native plants, and you may have to seek out nurseries that specialize in native plants to find species that are unique to your region. Keep in mind that nursery stock grown in your area will tend to be better adapted to local growing conditions. Also remember that when choosing exotic species, some can become invasive and take over existing native plant communities. Look for ways to naturalize these areas with native vegetation.

Naturalizing the Managed Landscape

Woods and wooded edges	Maintain undergrowth, small trees, shrubs, and leaf litter. Leave large wooded areas undisturbed whenever possible.
Old and open fields	Wait until after July 31st to mow. Plant warm-season grasses, such as big and little bluestem, Indian grass, and switchgrass to provide the best habitat. Changing mowing practices will allow for undisturbed nesting of grassland birds.

Unused lawn	Allow portions to grow taller and mow an attractive border along your wild edge, or naturalize with wildflowers or taller native grasses.
Clusters of trees	Plant flowering shrubs underneath or allow taller grasses to grow.
Fence rows and thickets	Hide unattractive fencing or increase privacy by planting climbing vines. Rather than removing weeds around fenced borders, let your fence line grow wild to provide a safe travel corridor and nesting sites.
Buildings	Plant native trees, shrubs, and flowers that produce seeds, berries, or nectar for birds around buildings. Evergreens can provide shelter and nesting sites.

Grasses planted at Cypress Knoll Golf Course, Palm Coast, Florida.
Courtesy of Audubon International.

Native grasses planted around pond at Huntsville Golf Club,
Shavertown, Pennsylvania. Courtesy of Audubon International.

A naturalized bank at Aurora Country Club, Aurora, Illinois.
Courtesy of Audubon International.

Native Trees and Shrubs for Wildlife

When choosing trees or shrubs, use native species that are well adapted to your area and your growing conditions. The following lists are adapted to broad regions across North America.

EASTERN REGION

TREES

American Beech (*Fagus grandifolia*)
Eastern Red Cedar (*Juniperus virginiana*)
Elms (*Ulmus americana*)
Hickory (*Carya spp.*)
Maple (*Acer spp.*)
Mountain Ash (*Sorbus docora*)
Magnolia, Southern (*Magnolia grandiflora*) and
Sweetbay (*M. virginiana*) - Mid-Atlantic and Southeast
Oak (*Quercus spp.*)
Pine (*Pinus spp.*)
Serviceberry (*Amelanchier canadensis*)
Sweetgum, American (*Liquidambar styraciflua*)
Tulip Tree (*Liriodendron tulipifera*)
Tupelo or Sour Gum (*Nyssa sylvatica*)

SHRUBS & SMALL TREES

Blueberry (*Vaccinum corymbosum*)
Dogwood (*Cornus spp.*)
Chokeberry (*Aronia arbutifolia*)
Hawthorn (*Crataegus spp.*)
Holly, American (*Ilex opaca*)
Sassafras (*Sassafras albidum*)
Spicebush (*Lindera benzoin*)
Viburnum, mapleleaf (*Viburnum acerifolium*)-Look for mapleleaf, nannyberry, arrowood, and blackhaw
Wax Myrtle or Bayberry (*Myrica pensylvanica*)
Winterberry (*Ilex verticillata*)

WESTERN REGION

TREES

Birch (*Betula spp.*)
Cascara Buckhorn (*Rhamnus purshiana*).
Cedar, Alaska Yellow (*Chamaecyparis nootkatensis*)
Cherry (*Prunus spp.*)
Dogwood (*Cornus occidentalis, C. pacific, C. California*)
Douglas Fir (*Pseudotsuga menziesii*)
Fir (*Abies spp.*)
Hemlock (*Tsuga spp.*)
Maple (*Acer spp.*)
Oak (*Quercus spp.*)
Pacific Madrone (*Arbutus menziesii*)
Pine (*Pinus spp.*)
Spruce (*Picea spp.*)
Western Red Cedar (*Thuja plicata*)

SHRUBS & SMALL TREES

Elderberry (*Sambucus spp.*)
Manzanita (*Arctostaphylos spp.*)
Red-flowering Current (*Ribes sanguineum*)
Red Osier Dogwood (*Cornus stolonifera*)
Salmonberry (*Rubus spectabilis*)
Serviceberry (*Amelnchier spp.*)
Snowberry (*Symphoricarpos spp.*)
Twinberry (*Lonicera involucrata*)

TEXAS & SOUTHWEST

TREES

Arizona Ash (*Fraxinus velutina*)
Acacia (*Acacia smalli, A. farnesiana*)
Ironwood (*Olneya tesota*)
Joshua Tree (*Yucca brevifolia*)
Mexican Elderberry (*Sambucus mexicana*)
Maple (*Acer spp.*)-Look for Southern Sugar Maple & Bigtooth Maple
Texas Madrone (*Abutus texensis*)
Pecan (*Carya illinoinensis*)
Palo Verde (*Cercidium floridum & C. microphyllum*)
Redbud (*Cercis canadensis & C. occidentalis*)
Brasil or Bluewood (*Condalia hookeri*)
Juniper (*Pinus spp.*)
Mesquite (*Prosopis spp.*)
Black Cherry (*Prunus serotina*)
Oak (*Quercus spp.*)
Desert Willow (*Chilopsis linearis*)

SHRUBS & SMALL TREES

Catclaw (*Acacia greggii*)
Fairy Duster (*Calliandra eriophylla*)
Hackberry (*Celtis spp.*)
Rabbitbrush (*Chrysothamnus nauseosus*)
Ocotillo (*Fouquieria slendens*)
Wolfberry (*Lycium pallidum*)
Golden Current (*Ribes aureum*)
Silver Buffaloberry (*Shepherdia argentia*)

References: *American Wildlife & Plants. A Guide to Wildlife Food Habits*, Martin et al., Dover Publications, 1951; *Landscape Restoration Handbook,* Harker et al., Lewis Publishers, 1993; *Landscaping for Desert Wildlife*, Arizona Game & Fish Department, 1992. *Landscaping with Native Plants of Texas and the Southwest*, Miller, Voyageur Press, 1991; *Plants for Wildlife in Western Washington*, Washington Department of Wildlife.

Developing a Plan of Action

Understanding what "naturalizing" your golf course means is a good beginning. But how do you figure out what kinds of plants you should or want to add, and where to put them; or what areas you simply want to stop maintaining; or what structural components, like nest boxes, would benefit wildlife on your own golf course?

1. Visualize the Golf Course Property

Your first step is to "see" your property as part of a larger landscape and then identify the various human and habitat components of the property. Think about your golf course not merely as the place where the game of golf is played, but rather as a natural setting in which the "playing surface" of the golf course is located. That is, your golf course is composed of a variety of areas, some of which are highly maintained and some of which may be minimally managed. In order to better visualize your property, we recommend drawing a map.

- Start with an outline of the whole golf course property. Identify the property around the golf course. Is it surrounded by undeveloped areas like forests and fields, or are you in a urban setting where human development surrounds the golf course?
- Draw in buildings, roads, and parking areas associated with the golf course.
- Draw in the "playing surface," including the tees, greens, fairways, and roughs.
- Draw in any special features or distinct habitats on your site. These could include standing bodies of water, such as lakes, ponds, and wetlands; moving bodies of water, like rivers, streams, and creeks; or woods or expanses of grasslands or meadows. Are there any places where you have identified specific wildlife activity?
- Draw in any "formal" gardens, or major plantings like shrubs and hedges. Include as much information about vegetation in and around the golf course as you can.

2. Assess the Golf Course Property

Once you have completed your map, you will need to ask yourself some questions about the golf course property, try to find out all the answers you can, and begin to document what you know by writing down what you have found out. The following information will guide you in making decisions appropriate for you, your golfers, your course, and wildlife.

- **Describe the location of your golf course.** In what state, county, city, or town is it located? In what plant zone, geographical region, or ecological region is it?
- **Describe the type of golf course.** Is it public or private? Is it part of a resort or residential community? How many members are there and how many rounds of play annually?
- **Describe the size of the golf course property.** How many total property acres are there? How many acres are under your management? How many acres is the golf course itself? How many holes?
- **Describe the types of wildlife on your course.** Do you keep an ongoing list of the species you've observed? If not, what resources exist in the area that might assist you in developing one (golfers or club members with wildlife interests or expertise; a local school, community college or university; local birding clubs, natural history groups, or local sporting clubs)?
- **Purchase a set of wildlife field guides** by authors such as Peterson or Stokes so you can determine and learn about the kinds of birds, mammals, reptiles, amphibians, insects, and plants found in your area. Good field guides have range maps to indicate where species are during all seasons of the year and what plants are appropriate for your temperature ranges.

3. Develop a Plan of Action

Set realistic goals and priorities for what you want to accomplish on the golf course. Take into consideration the expectations of the golfers that play your course. Remember that they may have conflicting ideas about the aesthetic vision of the golf course. Areas that you may decide to "naturalize" may look merely neglected to some golfers.

Develop a written plan of action, including:

- species of animals you are or will be managing,
- plants and their locations (include both specialty gardens and naturalized areas),
- structures or projects you will initiate and their location,
- an estimate of the time it will take to complete each project.

Remember that you don't have to change everything all at once. Don't take on too many projects too quickly, but work slowly to educate and explain to golfers what you are trying to accomplish.

Identify your own management restrictions or limitations, or potential problems such as slow play. If naturalizing areas of the golf course slows play, it is neither good for the game nor good for the ultimate success of your conservation efforts.

The more you understand about fundamental wildlife and habitat concepts, and the more experience you gain implementing basic wildlife management projects, the more knowledgeable you can be when educating and enlisting the help of golfers. These efforts will set the stage for your future actions and successes. Keep in mind the benefits of naturalizing your property:

- Reduced maintenance costs,
- Reduced water and chemical use,
- Enhanced visual diversity,
- Greater habitat diversity,
- Greater wildlife diversity.

Wildlife Management Projects

Before you begin your projects, remember that your ultimate goal is to increase habitat diversity and consequently enhance the habitat for wildlife. The best way to stay focused on this goal is to remember the basic requirements for wildlife survival—food, water, cover or shelter, and space. As you decide which habitat projects are most appropriate for your course, here are some additional considerations about providing for the needs of wildlife.

Food

When planning to attract wildlife, providing food is often the first activity that comes to mind. You can manipulate food resources by simply setting up a bird feeder or adopting a landscaping plan that includes plants of high food value to a variety of wildlife species (i.e., fruit-, nut-, or seed-bearing vegetation).

When you enhance food sources for wildlife, some things to consider are: the food preferences of different species, fruiting habits and seasonal availability of plant material, and selection of plants for food and cover. Some animals will eat a great variety of food items, whereas others eat only a few kinds of foods.

Insects are a vital food source for many birds and mammals. If you keep some areas unmowed or mow seasonally, you will not only cut down on maintenance costs, but you will also provide excellent areas for beneficial insect populations. Use caution when applying pesticides and seek the least toxic product that can achieve the desired results.

Water

The availability of water is often the most important factor in attracting wildlife. Birds use water not only for drinking, but to keep their feathers clean in order to retain body heat. Managing your water resources should be a full-year commit-

Buffer zones around water bodies at
Robert Trent Jones Golf Club, Manassas, Virginia.
Courtesy of Audubon International.

ment to enhance the value of your golf course for wildlife. Most golf courses incorporate either man-made or natural water sources into their design to add both beauty and challenge to the game, and both can be managed with wildlife needs in mind.

A primary consideration in managing water areas for wildlife and the game of golf is chemical use. Runoff from chemical applications can greatly affect the balance of plant and animal life in ponds, streams, and other water bodies. Adopting an Integrated Pest Management (IPM) program, creating "no spray" zones, and buffering the edges of water features with vegetation will help you minimize chemical inputs into water resources.

You can also manage streams, creeks, and rivers by planting the banks with a variety of plants to prevent erosion and to provide food and cover for wildlife. Again, runoff into these areas should be evaluated and monitored.

Another way to provide water for wildlife is to include a birdbath by the clubhouse, near bird feeders, or in open areas that have no water. Small ponds are also important wildlife features because of their shallow bottoms. Aquatic vegetation will provide cover for many insects, amphibians, and reptiles that inhabit the pond. This diversity of life will attract larger mammals and birds.

Marshes, swamps, and low boggy areas are best left alone and possibly incorporated into the design of your course. These areas provide valuable habitat for a variety of plant and animal life and may also be used by forest or meadow species.

A wood stork searches for food in a pond at PGA at the
Reserve, Port St. Lucie, Florida.

Shelter

Shelter or cover is a general term applied to habitat that provides protection
for animals to carry out life functions such as breeding, nesting, sleeping, resting,
feeding, and travel. Anticipating the need for cover is related to planning food
sources because animals often will not come to food if there is not a protected
place for them to eat it. Hedgerows and taller grasses will be used as safe travel
corridors by wildlife seeking food or water. Landscaping with a variety of flowers,
grasses, shrubs, and trees will help accommodate a variety of wildlife, from ground-
dwelling species to those that prefer living in the treetops. Forest understory also
provides cover for safe travel and nesting, and brush piles can be added to a
woodlot understory to enhance cover for small mammals.

Space

The area that you manage for wildlife may be small or it may include many
acres. An important component of habitat, space is the one over which you may
have the least control. However, when you manipulate the other habitat components—water, food, and cover or shelter—often the amount of space needed by
wildlife is not as critical a factor. You can manipulate or enhance the other habitat
components by landscaping with trees or shrubs of high wildlife food value,
mounting nest boxes, building ponds, or allowing taller grasses to grow. By imple-

menting these projects, you will reduce the amount of living space required by an individual animal and increase the chances of attracting wildlife to your property.

When you consider the kinds of wildlife you might attract to your course, remember to think beyond the boundaries of your property. Surrounding land use will have an impact on the types of wildlife you can attract. Even though you may not own or manage large acreage, the combination of your golf course and adjacent natural areas may add up to sizable habitat. On the other hand, if you are surrounded by housing or commercial development, you may not attract certain species to your property no matter how much you manipulate the habitat components. Keeping in mind this "big picture," the key to successful management for wildlife is properly using and enhancing the space you have.

When you provide food, water, and cover to attract wildlife, you must also consider the arrangement of these habitat elements. For instance, if you plant shrubs and trees to provide food and cover, locate them near a water source to complete habitat requirements. Keep in mind that the key to managing your property for wildlife is **diversity**. The more diverse the habitat on your course, the more species you will attract.

Suggested Wildlife Habitat Enhancement Projects

PLANT COMPONENTS

Incorporate native plantings. Trees and shrubs with edible berries, nuts, or seeds provide the best food source for wildlife in your region. Native plants will not only benefit the greatest variety of wildlife, they will tolerate the environmental stress of the region without the overuse of water and chemicals.

Plant vegetative layers. Incorporate tall grasses, small shrubs, small trees, and larger trees, developing vertical vegetative layers. Multi-level habitats will provide edges that will better meet the needs of a variety of wildlife.

Add shrubs, hedges, and trees. Planting shrubs and hedges provides shelter and nesting sites. Evergreens with dense or thorny branches are especially good. Mix groups of shrubs, trees, and taller grasses to create edge.

Include nectar sources. Plant flowers for hummingbirds and butterflies.

Leave woodlot understory. Reduce maintenance in out-of-play areas under trees that are presently being cleaned, manicured, or possibly even where turfgrass is growing.

Designate no-mow areas. Let grasses grow in out-of-play areas to create habitat for grassland nesting birds, to create buffer areas between fairways, and for contrast between the various areas of the golf course.

Establish vegetative buffers. Shoreline plants around pond or stream edges increase food and cover sites for wildlife. Add aquatic and shoreline plants to lakes and ponds.

STRUCTURAL COMPONENTS

Create a brush pile. This provides valuable shelter and basking sites for small animals.

Mount nest boxes. There are 86 species of birds that rely on cavities for nest sites.

Provide standing dead trees (snags). When you have the opportunity, leave snags standing, as long as they don't cause safety concerns. They are important as perches and nesting sites for a wide variety of wildlife.

Maintain a variety of bird feeders. Suet feeders, ground feeders, and hanging feeders will supplement seasonal food lapses and are especially helpful to wildlife in late winter and early spring when natural food supplies are lowest.

In addition to the suggested wildlife habitat enhancement project in which you might want to become involved, we have also included more detailed information about a number of these projects, including: Creating a Garden for Butterflies and Hummingbirds, Building a Brush Pile, Mounting and Monitoring Nest Boxes, Tree Snags, Feeding Birds at the Clubhouse, and Water Plantings for Wildlife.

Creating a Garden for Butterflies and Hummingbirds

Planting flowers that are attractive to people, as well as to hummingbirds and butterflies, can create an added dimension to areas around the clubhouse and provide an opportunity to educate your members and the public about the enjoyment and satisfaction of enhancing habitat for wildlife. For best results, your garden needs to get at least a half day of sunshine. A southern exposure is best, but gardens that face east or west will also work.

Planting flowers that will attract butterflies and
hummingbirds will also appeal to people.
Courtesy of Audubon International.

Hummingbirds are the smallest birds on earth, and measure about 3-1/2 inches long. They can fly backward, forward, sideways, and straight up and down. Because they move so quickly, they may only be visible when they stop to hover and feed. Hummingbirds have slender, pointed bills that are adapted especially for probing flowers for nectar.

Hummingbirds are attracted to a diversity of habitat from shrubs to trees. They do not normally nest in salt marshes or grasslands. They like both sun and shade. They need small branches with a view to overlook available flowers, look out for predators, and keep other hummers from their territory.

Hummingbird plantings can be a combination of wildflowers, perennials and annuals, planted in small gardens, containers, or hanging baskets. There should be overlapping bloom (from early spring to late fall) to provide a continuous supply of nectar. Gardens planted with low, medium, and tall plants provide open areas for nectar sipping. Plant tubular-shaped, single flowers (double flowers often do not have nectar). While there is some preference for red or

A wildlife garden at the Standard Club, Duluth, Georgia.
Courtesy of Audubon International.

orange flowers, birds find nectar in flowers of all colors. Because hummingbirds also eat insects and small spiders, do not use pesticides on your hummingbird plantings.

Butterflies are insects known for their large, usually bright-colored wings. Like other insects, they are cold-blooded. They rely on the sun to keep them warm and raise their metabolism so they can fly. On sunny days, they can be seen basking on light-colored rocks, which reflect more of the sun's light to their bodies, or sitting on flowers free from shade. Butterflies remain inactive on cloudy days.

All butterflies go through dramatic changes before becoming colorfully winged creatures. Their metamorphosis proceeds through four stages: 1) egg, 2) larva, or caterpillar, 3) pupa, or chrysalis, and 4) adult butterfly. To entice butterflies to your garden, you can grow plants for both adult butterflies and caterpillars to eat. You can also provide stones for sunning and a shallow source of water.

Plantings for a Wildlife Garden

Although there are many flowers that will provide nectar, fruit, or seed, the following table includes only those that are easily grown and require minimal maintenance. Because some plants are more easily obtained using the Latin name, it is included next to the common name.

Flower	Hummingbird	Butterfly	Description
Perennials			
Asters (*Aster* spp.)		✓	Pink, purple or white; fall bloom
Beebalm (*Monarda didyma*)	✓	✓	Red; summer to early fall
Bleeding heart (*Dicentra* spp.)	✓		pink; spring to fall
Butterfly weed (*Asclepias tuberosa*)		✓	orange; midsummer; a favorite nectar-producer
Black-eyed Susan (*Rudbeckia* spp.)		✓	yellow; August and September
Cardinal flower (*Lobelia cardinalis*)	✓		red; late spring through fall
Catnip (*Nepeta cataria*)	✓		purple; early summer
Columbine (*Aquilegia canadensis*)	✓		red; early spring through summer
Coreopsis (*Coreopsis* spp.)		✓	yellow and orange; summer to fall
Coral bells (*Heuchera sanguinea*)	✓		pink; late spring to summer
Fuchsia (*Fuchsia* spp.)	✓		many colors; early summer to fall; you'll likely have to buy this as a hanging basket
Liatris (*Liatris* spp.)	✓	✓	purple; summer to fall
Milkweed (*Asclepias* spp.)		✓	an essential "host" plant for monarch butterfly caterpillars

Annual Flowers

Cosmos (*Cosmos* spp.)	✓	shades of pink; late summer and fall; easy to grow
Calendula (*Calendula officinalis*)	✓	yellow and orange; blooms midsummer to fall; may reseed themselves
Marigold (*Tagetaes erecta*)	✓	only the single-flowered varieties produce nectar; get "lemon gem" or "tangerine gem"; blooms summer to fall; may be easier to start indoors
Nasturtium (*Tropaeolum majus*)	✓	orange, yellow, and red; summer; large seeds are easy for kids to plant
Scarlet runner bean (*Phaseolus coccineus*)	✓	red; summer bloom; needs poles or fence to climb on
Snapdragon (*Antirrhinum majus*)	✓	many colors; spring to fall
Sunflower (*Helianthus annuus*)	✓	yellow; August and September; giant gray stripe is best for seeds
Zinnia (*Zinnia elegans*)	✓	many colors; late summer to late fall

Vines *Ask for both of these by the Latin name to be sure you get the right vine. Other varieties can be invasive.*

Trumpet creeper (*Campsis radicans*)	✓	orange-red; midsummer to early fall; grows fast with little care; needs strong support
Trumpet honeysuckle (*Lonicera sempervirens*)	✓	red; spring through early fall; twining vine needs support

Building a Brush Pile

You can easily provide cover for small mammals and birds by building a brush pile on your property. In addition to providing wildlife shelter and protection, a brush pile is also an excellent way to make use of downed limbs, twigs, and debris.

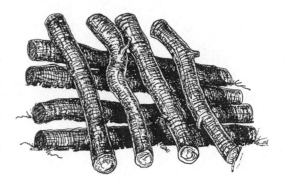

Bottom half of brush pile construction.

To build your own brush pile, lay four logs (6 feet long and 4 to 8 inches in diameter) parallel to one another about 8 to 12 inches apart on the ground. Then place four more logs of the same size across and perpendicular to the first four poles. These will keep "tunnels" open under the pile.

Next add brush: larger limbs first, then smaller branches, until you've created a structure 4-inch to 6-inches in height and diameter. Sticks and branches can then be continually added to the top as the pile rots at the bot-

Top half of brush pile construction.

The Country Club of Florida created log and brush piles for wildlife in out-of-play areas, Village of Golf, Florida. Courtesy of Audubon International.

Brush pile created at Tartan Park Golf Course, Lake Elmo, Minnesota.
Courtesy of Audubon International.

tom. This will provide food for an abundance of earthworms, enriching the soil and reducing the need for trash collection.

If you want a brush pile for birds to use, but not rabbits, pile brush one or two feet off the ground on cement blocks. It will no longer shelter rabbits.

Mounting and Monitoring Nest Boxes

Putting up nest boxes is an easy and economical way to increase bird nesting success and help balance the loss of natural habitat. The following information will provide you with the necessary details about cavity nesting birds, assessing habitat, and placing and monitoring nest boxes.

Although the majority of birds build their nests in the branches of shrubs or trees or hidden in taller grasses on the ground, certain species prefer to nest in hollow cavities. These birds are called **cavity-nesters.** Woodpeckers can make their own nest holes by excavating them in dead trees (snags). However, there are many other cavity-nesting birds, such as chickadees, bluebirds, and house wrens, that cannot make their own holes. These birds rely on old woodpecker holes and natural cavities in dying trees for nesting sites.

Unfortunately, the number of these natural nest sites has been steadily declining over the years. Many forests and farms where tree snags are common have been developed to make way for homes, businesses, schools, and shopping centers. And because dead trees are often considered "eye sores" or hazards, they are frequently removed by land managers.

A wood duck box was placed near water and cover at Eagle Knoll, Hartsburg, Missouri. Courtesy of Audubon International.

The reduction in the number of available nesting sites has increased competition among birds for the remaining sites. For example, the North American bluebird population declined by nearly 90% from 1930 to 1980. Loss of habitat and competition with house sparrows were significant causes of this decline. Fortunately, nest box placement for bluebirds has now begun to reverse the decline of bluebird populations.

Nest boxes are birdhouses that are specifically tailored for cavity-nesting birds. The size of the opening and the dimensions of the box have been calculated to fit the size of the bird and its nest. For example, the standard songbird box is sized to fit a number of small birds, but the entrance hole is too small for the aggressive European starling.

Putting up a nest box is an easy and economical way to increase nest sites for cavity-nesting birds and to help them achieve nesting success. To begin a nest box project on your golf course, you must determine how many boxes you want to put up, what type of birds you are likely to attract, and who will build, monitor, and maintain the boxes.

Start by surveying the types of habitat available at your golf course. For example, locate open areas, woods, and areas with water—such as ponds or streams—that won't interfere with the golf game and that are isolated from frequent human activity. After surveying your course's habitat, you can then identify which birds are likely to take up residence and how many boxes the course can support.

Habitat Preferences of Cavity-Nesting Birds

Habitat Type	Bird Species
open areas— lawn or field	bluebirds
	kestrels
wooded sites or edges	wrens
	chickadees
	tufted titmice
	nuthatches
areas with water	tree swallows
(pond, stream, or wetland)	violet green swallows
	purple martins
	wood ducks (wetland preferred)

Choose the Correct Nest Box. The basic songbird nest box will house all of the birds listed above except for wood ducks and kestrels. These larger birds need bigger boxes. If you choose to purchase a nest box, check the dimensions carefully to make sure it is built for the birds you want to attract. Many lawn and garden stores sell suitable nest boxes, as do bird specialty stores and mail order businesses. Be sure that you can open the box easily. This is essential for cleaning out old nests and monitoring the nesting birds.

Mount Your Nest Boxes. Follow these tips for successful nest box placement:

- Mount nest boxes on metal poles, trees, or fence posts. A metal post will discourage predators such as raccoons, cats, and snakes from climbing to the nest.
- Mount the boxes four to five feet above the ground. Keeping the boxes within easy reach will make checking and cleaning them much easier.
- The nest box opening may face in any direction, but positioning the entrance hole away from prevailing winds will help keep the nest dry.
- Most birds begin searching for nesting sites in February in the South and March in the North, so boxes should be up by early spring. Boxes that are placed by late spring are also valuable and may attract birds ready to begin their second brood.
- Consider the first year of your nest box project an experiment. Some boxes may go unused, while other boxes may be eagerly sought out by several birds. Try placing your boxes in a variety of locations to see which ones are most successful. As a general rule, place boxes away from areas of high traffic.

Also keep in mind that birds have different territorial requirements that affect the number of birds of the same species that will use a given area. Boxes can be

Nest Box Construction

For small songbirds, including bluebirds, wrens, swallows,
chickadees, nuthatches and titmice.

Materials

1. 1"x6", 60" long, sea-
   soned pine, spruce, cedar, redwood, or ¾" ply-
   wood.
2. Galvanized nails.
3. One 1 ½" wood screw.
4. Tools: Hand saw or
   electric saw, screw-
   driver, hammer, tape measure.

The finished
bluebird
nest box

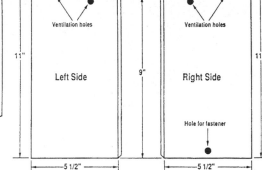

Construction Notes

1. Draw pieces out on board and cut them with
   a hand saw or electric saw.
2. Make entrance hole 1 ½" in diameter, 5 ½"
   from the floor on the front panel. For mountain and western bluebirds drill 1 ⁹/₁₆" entrance hole.
3. Drill two ¼"-diameter ventilation holes
   near the top of each side piece.
4. On right side panel, drill a ⅛"-diameter hole
   where it overlaps the floor for the fastener
   (the wood screw) that will hold the side-
   opening door closed.
5. Drill holes near the top and bottom of the
   back panel for mounting the box.
6. Nail the pieces together with galvanized
   nails. Fasten the right-hand side panel near
   the top only, about 8" up, with one nail
   through the front of the box and one nail
   through the back.
7. A small strip of wood can be nailed to the
   top of the roof along the back panel to keep
   out rain.
8. Stain house (outside only!) if you wish, but
   remember that birds prefer natural wood.
9. Do not add a perch to the box, as it will
   attract sparrows.

(Modified from: "The Bluebird Book," by Donald and Lillian Stokes. 1991.)

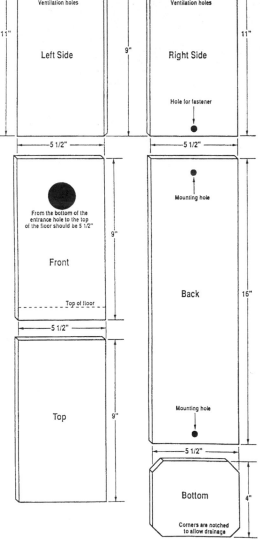

placed in pairs, 10 to 20 feet apart, to reduce nest site competition between different species.

Monitor Your Nest Boxes. Once you put up nest boxes, it is important to monitor them on a regular basis to help ensure nesting success. During the breeding season, from mid-March through August, open the box once a week or every two weeks. When you open the box, you can count the number of eggs or young, detect predator problems, and discard old nests once each group of young has left the box. Opening the boxes occasionally will not frighten the adult birds into abandoning the nest. Just open the box, take a quick look inside to count the number of eggs or young, check for disturbance, and close it again.

The only time to refrain from checking the boxes is after the young are about 12 days old. Opening the box during the 12- to 18-day period after birth may provoke the young into leaving the box too early. If you're not sure how old the young birds are, play it safe and refrain from checking the box for two weeks. Then look inside and remove the old nest if it is empty. Cleaning out the box will prevent a buildup of mites and other parasites that prey upon birds.

Record information about your nest boxes to learn more about the birds using them and keep track of the number of birds successfully fledged. Copy the nest box chart below or use a 5×7 note card for each box to record information about bird activity.

Chart to Record Bird Activity			
nest box #		birds first enter box	(date)
start of nest building	(date)	first egg laid	(date)
total number of eggs		first hatching	(date)
number of eggs hatched		number of young fledged	
comments/concerns/problems:			

Tree Snags

A dead tree plays a vital role in a woodland or forest community. If snags exist on your property, leave them standing when they pose no safety hazard and don't be concerned about "cleaning up" the forest floor. Standing or downed, dead trees provide an abundance of life for many years.

Cavity created in a snag at the Cranberry Resort Golf Course, Collingwood, Ontario, Canada. Courtesy of Audubon International.

To many wildlife species tree snags serve as an invaluable habitat component. Snags are used by nearly 75 species of birds and at least 50 mammal species in the United States. They are used as food sources, nesting sites, and perching sites, and for territorial establishment. Insect larvae commonly occur under the bark and in the soft wood of snags. The scarcity of nesting and roosting cavities is a major factor limiting the populations and diversity of many cavity-nesting birds. Primary cavity-nesting birds, like woodpeckers, excavate their own nest holes. Secondary cavity nesters, like bluebirds and nuthatches, use cavities previously excavated by another bird.

Background Information

A snag should be at least 6 inches in diameter and 15 feet tall. However, the larger the snag, the greater its value for wildlife. Small-diameter trees may be suitable for some kinds of birds, but they are not as useful for larger wildlife.

Snags are classified as either hard or soft snags. A hard snag is a dead or partially dead tree with at least some limbs remaining and sound wood. A soft snag is a standing dead tree in an advanced stage of decomposition with few if any limbs and advanced hard rot. Though they are excellent foraging areas, soft snags are neither as long-lasting nor as good for nesting habitat as are sturdier hard snags. However, both kinds are used by wildlife.

Not all snags are equal in quality or suitability for providing nesting habitat for birds. Hardwood snags are preferred over conifers. Basswood and birch are good cavity-producing trees, but most oaks (except black oak) are not. Pine and tamarack snags are long-lasting and will provide good nesting and perching sites for ospreys if they are adjacent to lakes.

Short-lived tree species such as aspen often provide suitable snags before the other forest trees mature. Trees infested with fungal heart rot often provide suitable cavities sooner than the natural death and decay process in sound trees. Trees notorious for heart rot, like beech, should be considered valuable as potential wildlife nest trees.

When snags fall down, they should be left on the ground to provide food and cover sites for birds, small mammals, reptiles, and amphibians. Logs can also provide drumming sites for ruffed grouse.

Benefits of Tree Snags

- Dying roots and fallen limbs and leaves contribute substantial organic matter and nutrients to forest soil. Living trees rely on these nutrients for growth.
- Fallen foliage and decaying food give life to the forest floor. Here an abundance of insects and macroinvertebrates find moisture and food. Woodland birds such as the wood thrush, ovenbird, and veery depend on these creatures for survival.
- Mosses, mushrooms, and microscopic bacteria on a dead log help break up dead organic matter. This continuing process creates fertile forest soil.
- Woodpeckers drill into dead trees to get at insect larva that burrow beneath the bark. Twenty-one species of woodpeckers live in North America. They nest in tree cavities and leave valuable holes behind for the other 65 species of North American cavity-nesting birds.
- Large cavities that form when a tree's inner heartwood decays provide nesting sites and protection from weather and predators. Raccoons, squirrels, and owls are some of the species that regularly use these larger holes.
- Hawks, owls, and other birds of prey perch on the upper branches of snags to scout surrounding territory for a meal. Lack of foliage offers a clear view for hunting.

Feeding Birds at the Clubhouse

Adopting a bird feeding program at your clubhouse is an enjoyable way of attracting wildlife to your golf course and providing your members with a first-hand view of the wildlife living on and around your property. A successful bird feeding program is one where the habitat requirements of food, water, and shelter or cover for birds are met at your clubhouse. A quick look at how birds survive in the wild should be helpful so that you can best satisfy their needs.

Birds spend most of their day foraging for food. They are remarkably adept at finding a diversity of food items. Most birds spend the winter or nonbreeding time in one particular area or territory that contains reliable food sources. Your bird feeding goal is to have your golf course on their list of daily stops. It is important to have your feeders in place by mid-fall because this is when feeding territories have already formed. Although it may seem as if birds are spending their entire day at your feeders, they actually spend a significant amount of time foraging for wild foods. Recognize that each species prefers a different type of food and likes feeding in different locations. The closer you can duplicate these preferences, the better your feeding program will be.

Cover is very important for protection from adverse weather and predators. Many birds prefer the **edge** of a habitat—the zone where two kinds of vegetation meet. When you are trying to determine where to place your feeding station, try to locate the feeders in sight of the clubhouse, but also near some shrubs or trees. Providing clean, open water at your feeding station will help you attract more birds. Water is important not just for drinking, but also for bathing. In general, birds prefer water on or near the ground.

A well-rounded feeding program should include high-energy grains, seeds, and suet. Variety is the key. The greater the variety of foods offered, the more species of birds you will attract to your yard. You will be successful, however, if you simply select black-oil sunflower seeds, white-proso millet (especially for the ground-feeding birds), and suet. Suet may be purchased in specialty bird stores in a variety of mixes that are highly nutritious and weather-resistant. Stay away from seed mixes that have a high percentage of useless fillers—milo, flax, wheat, red millet, or oats. These seeds will not be eaten and only waste money. Several specialty foods are very attractive to birds. Add niger (thistle), safflower, peanuts (whole or pieces), or a mix of the above at your feeder to attract many different kinds of birds.

We recommend that you find a local or mail order company that specializes in wild bird feeding. They will be more likely than grocery or hardware stores to carry high-quality bird seed and varieties of suet, as well as offering bird feeders, nesting boxes, binoculars, books, and other educational information specifically geared to people who want to start bird feeding programs.

How food is presented is nearly as important as the type of food offered. Start with a tray or platform feeder—either hung or set on a pole. There are feeders available that will last for many years and blend into the landscape and architecture of your clubhouse. We recommend placing your feeding station near a building or your seed source so that filling the feeders each morning takes just a matter of a few minutes.

Several types of feeders at different heights will cater to different types of birds. Make sure your feeders keep the food dry and have adequate drainage. If possible, place feeders in protected southern exposures for winter feeding and semi-shade in the hot months. For aesthetic reasons and the health of the birds, it is vital that you clean up hulls or leftover seed periodically to help prevent molds, *Salmonella,* and other diseases from developing in the litter.

A good way to let your members know that you are concerned about and interested in wildlife is to provide an opportunity for them to record their observations of the birds that visit your feeding stations, as well as birds they may see while they are out golfing. You may find that you already have some expert birdwatchers in your membership who might volunteer to help with this effort.

Types of Feeders	
hanging tube feeder	Easy to fill and food is clearly displayed to birds. Metal-reinforced perches and holes help prevent squirrels from monopolizing hanging feeders.
house-style feeders	Large models hold a lot of seed. Use a baffle on the pole and place the feeder away from hanging branches if you don't want squirrels jumping from trees or climbing the pole and eating all the birdseed.
ground feeders (open tray or scatter seed directly on the ground)	Attract cardinals, mourning doves, dark-eyed juncos, white-crowned sparrows, and other birds who prefer to feed on the ground.
suet feeders	Provide high-energy fat for birds, including woodpeckers, nuthatches, chickadees, and titmice. Suet is low-maintenance, lasts a long time, and can be hung from onion bags or placed in a wire cage.

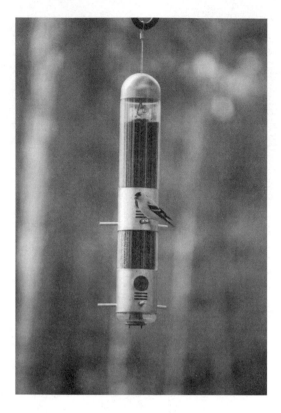

Hanging tube feeder. Credit: Chris and Ellie McIntire.
Courtesy of Audubon International.

House-style feeder. Credit: Chris and Ellie McIntire.
Courtesy of Audubon International.

Water Plantings for Wildlife

Birds and mammals rely on ponds and other water sources to drink, bathe, and cool off, while numerous species of fish, salamanders, frogs, and aquatic insects live or fulfill vital aspects of their life cycle in the water.

A pond's attractiveness to wildlife is largely dependent on the number and variety of aquatic and shoreline plants surrounding it. While you may be familiar with planting trees, shrubs, or flowers on your property to attract wildlife, planting a pond requires special consideration and generally presents even experienced gardeners with unique challenges. Network with other golf professionals who have been successful and learn from their experiences.

To get started, you need to gather some basic information about your pond so that you can choose plants and other enhancement projects that are best suited for your site. Your initial assessment should include:

Aquatic Vegetation

size
Make an estimate of the size of your pond. If it is very large and currently has limited vegetation around it, you may want to start by planting a small portion. This will allow you to "get your feet wet" without taking on more than you can handle.

sunlight and water depth
Use a yardstick to measure the depth to the bottom along the edge of the pond. Also note how many hours of sun the pond receives each day. Knowing this information will help you decide what plants will grow best in and around the pond. For example, many aquatic plants, including pond lilies, prefer full sun, while others, such as tall scouring-rush, can tolerate shade. Aquatic plants are also adapted to survive at varying water depths. Emergent plants, including arrowheads, bulrushes, and rice cutgrass need to be planted in the water to survive. Others prefer to be along the edge of the pond and can only tolerate moist soil, not inundation.

existing vegetation
Take stock of the existing plants in and around your pond. This will give you valuable information about what types of plants seem to grow well. In some cases, you may choose to expand existing natural areas or add to their diversity by introducing different types of shoreline plants.

wildlife value and aesthetics

While certain plants provide both food and cover, many offer only one or the other. Choosing native plants with a high wildlife value will help you attract the greatest diversity of species. At the same time, you may also want to choose a few plants simply for their aesthetic value. For example, blue flag iris and marsh marigold offer limited food and cover, but a small clump can provide beautiful ornamental interest along the shoreline of your pond.

Hand-planted cattails at the Carolina National Golf Club, Supply, North Carolina. Courtesy of Audubon International.

Pickerel weed in constructed wetland at the Carolina National Golf Club, Supply, North Carolina. Courtesy of Audubon International.

Hardy Shallow Water Plants

Emergent plants grow best in shallow water and prefer wet conditions. Plant in 6 to 12 inches of water. Height ranges from 2 to 4 feet.

1. Common Cattail
Typha latifolia

ZONE 3 • Leafy foliage, all season interest, and excellent wildlife food and cover value. A perfect tall background plant. 4' to 6' tall.

2. Arrow Arum
Peltandra virginica

ZONE 5 • Attractive glossy, dark green arrow-shaped foliage. Will tolerate some shade. Wood ducks and waterfowl eat the seeds. 2' tall.

3. Duck Potato/Arrowhead
Sagittaria latifolia

ZONE 4 • Strong, wide arrow-shaped foliage. White flowers with bright yellow centers. Underground tuber provides a water bird food source. 2' tall.

4. Pickeralweed
Pontederia cordata

ZONE 3 • Slick, broad, dark green foliage and bright lavender flowers. Easy to grow. Seeds eaten by waterfowl. 1' to 2' tall.

5. Bulrush
Scirpus sp.

ZONE 5 • Lush, dark green foliage. Water birds and song birds eat the seeds; stems and rhizomes are eaten by muskrats and geese; birds also nest in the upright stems. Valuable for controlling shoreline erosion. 3' tall.

6. Sedges
Carex sp.

ZONE 5 • Narrow foliage form dense clumps with uniform stature. Provides habitat for upland ground birds, mammals, songbirds, waterfowl and shorebirds. 1' tall.

7. Common Rush
Juncus effusus

ZONE 5 • Dark green spiky stems. Provides food, cover, and nesting for waterfowl. 2' to 3' tall.

8. Blue Flag Iris
Iris versicolor

ZONE 4 • Delicate violet blossoms and sword-like foliage. Plant near water's edge. 0 to 4" deep

9. Sweet Flag
Acorus americanus

ZONE 4 • Slender, upright with narrow sword-like leaves, light yellow-green flowers. Seeds eaten by wooducks, stems eaten by muskrats. 1' to 4' tall.

10. Marsh Marigold
Caltha palustris

ZONE 3 • Golden yellow flowers in early spring. Prefers cold climate, and light shade or cool running water. 1' tall.

Border Shrubs

Border shrubs can be planted along the borders of streams, ponds, or wetlands to add diversity and height to existing habitat. The following provide good food and cover for birds and can tolerate a variety of wet conditions and periodic flooding. Plant them above the water's edge. Height ranges from 5 feet to 20 feet.

common buttonbush (*Cephalanthus occidentalis*)	This food source for waterfowl also bears attractive flowers used by ruby-throated hummingbirds.
common winterberry (*Ilex verticulata*)	Berries provide a winter food source for birds.
northern bayberry (*Myrica pennsylvanica*)	Provides both cover for nesting sites and food for many songbirds.
pussy willow (*Salix discolor*)	Grouse eat buds, American goldfinches use for nesting, and mammals and songbirds eat the showy fruits.
red chokeberry (*Aronia arbutifolia*)	Berries are eaten by 12 species of song birds; provides fall color interest as well.
shadblow serviceberry (*Amelanchier canadensis*)	Berry food source for 36 species of songbirds.
silky dogwood (*Cornus amomum*)	Cover, nesting sites, and food source for birds; adds fall color interest.
speckled alder (*Alnus rugosa*)	Provides food for 15 species of songbirds, including goldfinches and pine siskins.

10 *Wildlife on Golf Courses*

One very important aspect of developing your plan of action and succeeding with your enhancement projects is understanding as much as you can about the wildlife that inhabits your golf course and their basic requirements. There are several species that are common to golf courses for which you may wish to enhance the golf course habitat. There are also some species that create problems for many golf course managers. We have included some basic information about those species and encourage you to learn more about them as well as other species that inhabit your course. Included are descriptions of bluebirds, turtles, bats, frogs, white-tailed deer, and Canada geese. There may also be times when wildlife pose a conflict with golfers and the game of golf, so also included in this section are some suggestions about how to deal with wildlife problems.

Bluebirds

A gentle bird with brilliant blue plumage, the bluebird has endeared itself to people across the country. Because of their early return from wintering grounds each year, bluebirds have become a symbol of spring. There are three species of bluebirds in North America: eastern, western, and mountain. Differing mainly in feather color and territory, these bluebirds are all likely to be found in largely open habitats such as meadows, parks, orchards, and farm fields.

Though once commonly sighted, bluebirds have dramatically declined in number over the past 50 years. This decline has been caused primarily by loss of habitat as development has spread to the countryside, and abandoned farm fields have reverted to woodland. Bluebirds nest in tree cavities, and loss or destruction of old, decaying trees has reduced suitable nesting sites and increased competition with other cavity-nesting birds. In addition, pesticide use has also taken its toll. When insecticides are passed through the food chain from insects to bluebirds, the toxic effects are often suffered by bluebirds.

Bluebirds are insectivores; they feed on a large number of insects, including snails, cutworms, and grasshoppers. An exposed post, wire, or branch is used as

Bluebirds exploring cavity nesting boxes.
Courtesy of Audubon International.

a perch for scanning the ground for insects. The bluebird will fly down to catch its prey and then return to the original or a nearby perch to eat. In winter, bluebirds are dependent upon berries to supplement their diet.

Courtship and nesting begins early in spring. The male chooses several potential nest sites, and with singing and tail and wing displays, urges the female to choose one. Nest building may start right away, or the pair may wait as long as six weeks before weaving the nest of grasses and sedges. Though the female builds the nest almost entirely by herself, the male will accompany her while she collects the nesting material.

Once the nest is finished, the female will lay three to five blue eggs, one each day until the "clutch" is complete. Incubation by the female then begins, and in 13 to 14 days the young hatch. The nestlings are fed insects and cared for by both parents until they are ready to leave the nest 17 to 18 days later. The family stays together for another two to three weeks, and then, if conditions are right, the adults will begin a second brood.

Though populations of bluebirds have declined, you have a good chance of attracting them to your property by providing bird houses designed specifically for bluebirds. These nest boxes are readily used by bluebirds, who are eagerly scouting for nesting cavities. Ideal sites to put up nest boxes include wooded clearings or edges, open areas with scattered trees, orchards, and fields with posts or wires for perching.

> For more information about bluebirds, contact the North American Bluebird Society, P.O. Box 74, Darlington, WI 53530-0074, (608) 329-6403, http://www.cobleskill.edu/nabs

Frogs

Like the canary in the coal mine, frogs are said to be good indicators of environmental health. There are at least two reasons why this is true. The first is that, instead of drinking like we do, frogs absorb water through their skin. That means they also absorb any kind of pollution that enters the water from the air or the ground, like acid in rain or pesticides in surface runoff.

The second reason frogs are good environmental indicators is that they spend part of their life in the water and part on land. As adults, most frogs live in damp places in the woods or near streams or ponds. But when mating season comes, usually in the spring, they migrate to ponds to lay their eggs. The frog calls that you hear in the spring are actually the attempts of males to attract females. Frog

The American toad. Courtesy of Audubon International.

eggs hatch into tadpoles, a completely aquatic stage that breathes with gills and eats algae. Depending on the species, they remain in the tadpole stage for as long as a year before they develop legs and lungs and move onto land as adults. This complex life cycle means that frogs are exposed to both water and land pollution during their lives, so they will die out if either is environmentally unhealthy.

This sensitivity to environmental health is one good reason to encourage frogs on the golf course. An abundance of frogs is strong evidence that you are taking good care of both the land and the water. Another good reason to want frogs around is that they eat lots of insects. One recent study found that a healthy frog population was removing over 50,000 insects per acre per year from the area under study.

Besides following good basic management practices to eliminate pollution and chemical waste, there are three basic things you can do to make sure your facility is inviting to frogs:

1. **Make sure there are breeding sites available in the spring.** Most frogs prefer shallow ponds (less than 3 feet deep) with emergent aquatic vegetation growing in them. Because fish eat tadpoles, the best breeding sites are ponds without fish, or at least ponds with extensive marshy areas too shallow for fish.
2. **Make sure there are good habitats for adult frogs.** Large patches of woods (or other native vegetation) located within 100 yards of the breeding site are best. These areas should be allowed to grow naturally, with good undergrowth and leaf litter, and dead sticks and logs should be left on the ground to serve as hiding places.
3. **Provide safe corridors between the woods and the pond.** Ideally the pond should have a no-mow, no-spray buffer zone all the way around it, and a similar corridor connecting the pond to the woods. These areas can be mowed once a year in late fall to keep them from getting too overgrown. Make them as wide as possible for the site.

There are close to 100 species of frogs in North America, so which species you have will depend on where you are. In general, there are three main groups that you are likely to see in most places. The first is the water frogs. These tend to be large and green and found near the water. Some, like the bullfrog, stay in ponds all summer, while others prefer dry land. They have smooth skin, but lack sticky pads on their toes. The second common group is the toads. They tend to be brown, dry, and warty, and can be found hopping around in broad daylight (unlike most frogs, which are nocturnal). The third common group is the tree frogs. They tend to be small with smooth skin, and they range in color from green to brown and gray. Tree frogs spend most of their time in the woods, but are frequently seen in the spring at breeding time. They can be distinguished by the large sticky toe pads, which they use to climb.

Turtles

Because they are cold-blooded, turtles are often observed basking on logs and alongside ponds, trying to warm themselves in the sun. Turtles have inhabited the planet for 185 million years and are unique in that they are one of the few vertebrates with a shell. Approximately 49 species of turtles occur in the United States, including sea turtles, freshwater turtles, and tortoises, who spend most of their lives on land.

Like all animals, turtles need food, cover, and water to survive. They can be found in a variety of areas, including ponds, lakes, wetlands, rivers, wet meadows, sloughs, bayous, and lagoons. Freshwater turtles are primarily carnivorous and feed on aquatic insects, crayfish, snails, leeches, fish, and other small animals. Plants too are an important part of their diet. Many turtles eat algae and the stems and leaves of several aquatic plants.

Though abundant aquatic vegetation or brush and tree trunks may attract turtles to a water source, the type of pond bottom and depth of water will affect which turtles inhabit a site. Many common turtles, including the stinkpot, snapping, and painted turtles, prefer shallow areas and soft, muddy bottoms. Others thrive where sand or gravel are the predominant substrate. The majority of turtles avoid fast-moving water.

Adjacent areas that can be used for nesting sites are also critical to turtle habitat. Some turtles nest on the sandy, muddy, or gravelly shores of the waters in which they live. Others, such as the Blanding's turtle, may nest up to one-half

A snapping turtle crossing turfgrass. Courtesy of Audubon International.

Turtle Habitat

Turtle	Range within United States and Canada	Habitat
painted turtle—most common and conspicuous of the basking turtles; often seen stacked on one another while sunning	coast to coast in the northern states and as far south as Louisiana	slow-moving, shallow water, soft bottom, basking sites, and aquatic vegetation
musk turtle "stinkpot"—named for their musky odor, sometimes bask in small trees	from New England to Florida and west to Texas and Wisconsin	shallow, clear lakes and rivers with slow current and soft bottom
pond slider—identified by bright patch of red or yellow on the side of the head	from southeastern Virginia south to northern Florida and west to Kansas, Oklahoma, and New Mexico	quiet ponds, streams, lakes or marshes with soft bottoms, an abundance of aquatic vegetation, and suitable basking sites
spiny softshell—looks like a giant pancake; powerful swimmers and can run well on land; handle with care—they bite	Great Lakes to the Gulf Coast	rivers with aquatic vegetation and soft bottom; fallen trees with spreading underwater limbs are often used for basking

Sources: *Turtles of the United States*, by Carl H. Ernst and Roger W. Barbour, 1972, and *Peterson First Guide to Reptiles and Amphibians*, by Roger Conant, Robert Stebbins, and Joseph Collins, 1992.

mile away from the wetlands and ponds where they reside during the warmer months. In winter, turtles need hibernation sites. Most burrow into muddy areas to wait out the colder months of the year.

To provide habitat for turtles on your property, you have to have an area of open water or wetland. Turtles will not be attracted to sterile ponds with very little vegetation or ponds where algaecides are frequently used. Vegetation both around and within the water is necessary for cover. A few logs placed within the water will provide an area for turtles to bask in the sun and warm themselves. Water 2 to 4 feet deep with a soft mud bottom will further help attract turtles. Adjacent shoreline that can be used for nesting and hibernation with minimal disturbance is also vital.

Turtles will eventually find their way to compatible habitat, especially if they already reside in nearby areas. *If you believe you have suitable habitat for turtles and would like to encourage their presence, it is important that you attract them naturally by providing appropriate habitat, rather than by introducing them.* Introduced turtles can spread disease and may have difficulty adjusting to an unfamiliar area. Learning more about which turtles occur in your region and their habitat needs will not only expand your appreciation for these wonderful creatures, but will also increase your chances of protecting and enhancing turtle territory.

Bats

There are 40 species of bats in North America, and none deserve their negative reputation for getting tangled in people's hair, drinking blood, or always carrying rabies. In fact, bats can be good neighbors and a vital resource for controlling pests and pollinating flowers.

Bats are furred, warm-blooded mammals with body lengths of 3 to 6 inches and wingspans varying from 8 to 16 inches. Most bats hunt flying insects and navigate by emitting pulses of sound through the mouth. Their sensitive ears hear the echoes reflected from even tiny insects. This allows them to steer toward prey and avoid obstacles. Bats also have keen eyesight, on which they rely for long-distance orientation.

Bats in North America eat primarily insects, such as cutworms, corn borer moths, potato beetles, and mosquitoes. A single bat can consume between 500 and 1,000 mosquitoes and insects in an hour, depending on the species and the size of the bat. Given this appetite, you can easily see why bats are the most important natural controller of insect pests that fly at night. Having a population of bats on your golf course can be a welcome addition to your integrated pest management program.

Unfortunately, nearly 40% of America's bats are on the Federal Endangered Species List or are candidates for it. Many factors have led to the decline of bat

Bat. Credit: M.D. Tuttle © 1989.

populations. When old buildings and barns are demolished, valuable bat roosting habitats are destroyed as well. Insecticides and pesticides are easily ingested by these insect-eating mammals. The popularity of spelunking or "caving" often puts people in bat caves just as young bats are maturing. Often if adult bats are disturbed by humans, they will abandon their young. Because bats usually raise only one pup each year, their populations do not increase quickly. Lastly, the myths about bats do not endear them to the general population.

You can help to ensure the survival of bat species in your area by: 1) supporting bat conservation efforts to protect existing natural nest sites and, 2) mounting "bat boxes" to provide additional nesting and roosting sites. A bat box is a simple wooden structure, much like a bird nesting box. It can be placed in a variety of locations, but bats prefer sites that are within a quarter mile of streams, lakes, or wetlands. Bat houses are used for nursery colonies, bachelor colonies, and hibernation.

Bat houses are currently a part of habitat enhancement projects on state parks, golf courses, farms, schoolyards, and backyards throughout the country. To allay any fears, be sure to educate golfers about the addition of bat houses on the course. Post bat house information or use your newsletter to explain this project. People generally welcome bats when they know that bats will be a valuable part of your pest management plan.

If you have bats on your property or like more information about bats, contact Bat Conservation International, Inc. P.O. Box 162603, Austin, TX 78716. (512) 327-9721, http://www.batcon.org

White-Tailed Deer

The population of white-tailed deer has been high in recent years, and many golf courses in urban, suburban, and rural areas provide perfect habitat. The overpopulation has caused golf professionals to seek out ways to minimize the damage deer can do to landscaping around the golf course. Understanding the white-tailed deer and knowing about their habitat needs may help reduce damage caused by large deer populations.

White-tailed deer show a preference for timbered areas, but are frequently sighted at edge habitat. They are commonly seen in woodland clearings, crop fields lying next to brushy and wooded areas, swamp borders, fencerows, and windbreaks.

Deer are browsers, meaning that a high part of their diet consists of twigs and buds from woody plants. This type of forage is especially important in winter, when grasses, herbs, and fallow croplands are covered with snow. At the appropriate season they readily eat corn, alfalfa, apples, and garden greens.

Deer are considered to be at their prime between 2-1/2 and 7-1/2 years of age. However, life expectancy is usually less than four years, especially where hunting pressure is high. Since big predatory animals such as the wolf and mountain lion have been pushed out of their former ranges, men, dogs, coyotes, and automobiles pose the most serious threat to deer today. They are also subject to certain bacterial and viral diseases.

Alamana Golf Course, Alamana, Iowa.
Courtesy of Audubon International.

To appropriately manage deer on your property, you must first survey your land. What conditions favor or discourage deer? Where do they spend the winter? What and where are they eating? What aesthetic and economic conditions affect how much damage from deer can be tolerated? In many areas, deer have become a problem when increasing development reduces the habitat size available to them. When deer are forced to live with more limited resources, they tend to strip vegetation and damage crops or landscaped trees and shrubs. In suburban areas, deer have few predators to reduce their numbers.

Striking a balance between deer and human needs is often the most effective way to manage deer. In recent years, landscape managers have used a variety of techniques to reduce deer damage including the following methods.

Repellents

- Human Hair—Collect human hair at a barber shop and spread it around specific plants or hang it in bags from branches.
- Soap—Drill holes in small bars of deodorant soap and hang them on the branches of valuable plants. (Leaving the wrapper on the soap makes it last longer.)
- Homemade Sprays—Some people have found homemade "hot sauce" sprayed on plants to be partially effective in repelling deer.
- Commercial Products—There are a variety of taste and scent repellents available at garden and hardware stores, and through catalogs.

Barriers

Valuable trees and gardens can be protected with an assortment of fences and aviary netting. Electric fences around plant nurseries, welded wire fences around some freestanding trees, and plastic fencing in some wooded areas can be effective. Also available is a dark mesh polypropylene material that can be strung from tree to tree around a property or planting; users say it's less obtrusive than regular fencing.

Canada Geese

Canada geese are perceived by many golfers and golf course managers to be a nuisance on the golf course. Their feces can leave a substantial mess that must be cleaned up regularly, and they are not easily deterred from going about their business. Many courses encourage geese when they first arrive, only to find that after several years of nesting success, they can no longer tolerate the large flock that has somehow made the course home. Without realizing it, many courses provide ideal geese habitat. Open water, an extensive food supply, and lots of open space is precisely what geese are looking for. Again, the more information you have at your disposal regarding a wildlife species and its habitat requirements, the better the opportunity you will have to resolve conflicts.

The long V-shaped flock and distinctive honking of Canada geese have long been synonymous with the coming of spring and fall. Canada geese are well recognized throughout the country, and they are becoming semidomesticated in urban parks and on golf courses with lakes, ponds, or wetlands. Here they find plenty of water, food, and nesting sites that are relatively safe from predators.

Around the turn of the century, populations of Canada geese were drastically reduced by unrestricted hunting on wintering grounds and migration routes. Today, geese are protected by law, and it is illegal to destroy nests, eggs, young, or adults without special state permission.

In recent years, wildlife biologists have identified a change in Canada goose populations and migration patterns. Many geese are no longer migrating great distances, but are forming "resident" populations that remain within a limited geographic area. Of concern are the dwindling numbers of Canada geese that breed in the arctic and subarctic and winter throughout the United States. Why these changes are taking place is not well understood and more research needs to be done. Loss of habitat, agricultural practices, and altered natural environments may all be linked to changing population dynamics of Canada geese.

Canada geese are grazing birds that feed on both wild and cultivated plants. They eat the rhizomes, roots, shoots, stems, blades, and seeds of grass and sedges, grain, bulbs, and berries. They also eat insects and aquatic invertebrates. Geese often spend the winter in agricultural areas, where they feed on postharvested grain and foliage.

Geese nest in a wide variety of habitats, but prefer sites with an open view—open fields near water, on islands, rocky cliffs, even in large tree cavities or artificial nest structures such as old tires. On the golf course, manicured water features provide a good view for nesting geese, while grass is plentiful forage. Geese usually mate for the first time in their second or third year, and pairs remain together as long as both are alive and healthy. They show strong nest site fidelity, and often return to the same spot year after year if they are successful in raising young. Geese raise an average of four to seven goslings each year.

Because of their size, intelligence, and wariness, geese are less subject to predation than most other waterfowl. Hawks and owls prey on immatures and some adults, and snapping turtles, snakes, and land-based predators take goslings that stray from parental protection. Parasites, diseases, and accidents also take their toll. The lifespan of Canada geese is potentially 15 to 20 years, though on average geese live about seven to nine years.

There are no easy solutions to geese problems on the golf course. If you are experiencing difficulties, following are some suggestions to help you assess your situation and determine the best solution. Keep in mind that Canada geese are protected by law. It is illegal to harm them in any way!

- Try to define the problem specifically. Are you dealing with resident geese or with a migratory population? Your method of control will be most effective if you define the problem, narrow down zones of damage, and effectively communicate with golfers.
- Survey your course to determine how you might actually be *encouraging* geese. Where are geese problems most prevalent? Do these places have open water that borders an in-play area? Do you mow grass right down to the shoreline, thus providing geese with the view they prefer? If the answer is "yes," then you are providing great geese habitat.

There are things you can do to try to lessen the impact of geese and discourage them from overrunning the course. Be realistic in your attempts to control geese. You will have much greater success if you try to strike a balance between Canada geese and the needs of golfers and maintenance personnel.

Choose the most effective control measure for your situation. Control measures fall roughly into three categories: **barrier methods, scare tactics,** and **intervention.** Some combination of these may prove most successful. Remember, the easiest time to deal with geese is before they start nesting.

Barriers include stringing monofilament line or wire around the edge of your ponds where geese are most prevalent. The string will not interfere with golf play, but will deter geese from easily walking onto the course. Planting native aquatic vegetation around lake and pond margins rather than mowing right to the water's edge helps to disrupt the expansive view, acts as a physical barrier, and provides excellent habitat for other more desirable wading birds and wildlife.

Scare tactics, including explosive noises, have been tried with very limited success. Remember, geese are big birds that don't scare easily. Mylar bird scare tape can be tied to stakes near ponds, but is of limited value for geese control. Dogs have been successfully used by golf course managers to chase or stalk the birds. Geese must be stalked in the early morning or evening until they feel threatened for their safety and leave.

Intervention is necessary when other control methods have failed and golfers will no longer tolerate geese on the course. At that point, local authorities should be contacted to help you deal with Canada geese. Each state has different regulations regarding goose control and your local department of wildlife or environmental conservation can help you determine the best method of intervention. This may include permits to disrupt nests or eggs, physical relocation (trap and transfer), or hunting.

Never destroy Canada geese without a permit. Canada geese are protected by law. Not only may your course pay large fines for killing geese, the resulting negative publicity can be devastating for your course and your profession.

Finally, recognize that you are not alone. Many golf course managers are struggling with the same problem. Encourage golfing organizations to which you belong to open a dialogue with the U.S. Department of Fish & Wildlife. Wildlife biologists are conducting ongoing research about Canada geese. They are interested in learning more about why migratory populations are declining while resident populations are on the rise. Communicating your concerns and observations can result in positive progress toward striking a balance between golf and geese.

Dealing With Wildlife Problems

The first step in dealing with wildlife conflicts effectively is to understand a wildlife species and its habitat. Learn about the species' preferences for food, cover, and water sources, as well as breeding requirements and territory size. When you learn more about the habitat and behavior of a "problem" wildlife species, you can identify a variety of alternatives that may correct the problem or help you discover the underlying conditions that caused the problem.

Be specific in evaluating and defining the problem. The more specific you are in defining the problem, the easier it will be to narrow down the solutions. Define what the problem is, where it's occurring, and when it occurs, as well as why it is occurring.

Review your management practices. Survey the property to determine whether there is something you are doing that causes or worsens the problem. For example, if you're eliminating understory vegetation that might support natural predators or removing shoreline vegetation around ponds, you would be creating a more positive habitat for geese.

Review solutions and choose the most effective alternatives. Consider your needs, the nature of the wildlife species, and your property. Also consider short- and long-term effectiveness. For example, you could increase vegetation around ponds to deter geese from the water source, promote the growth of understory, or get a dog to discourage geese from breeding on the property.

Share your knowledge about the wildlife species and the nature of the problem with golfers. People are apt to support your efforts when they know you fully understand the problem and are taking action to solve it.

The Right Thing to Do

Over the past ten years, I've traveled across the country and internationally. I've visited hundreds of golf courses and talked with golf course superintendents and managers, architects and designers, and golf professionals. When we talked about how a golf course should look or how it should play, everyone had an opinion. Like all really good discussions, we raised more questions than we had answers and the exploration frequently led us to essential questions about the game of golf.

How did we arrive at the current vision of a golf course? How did we come from a game played in linksland, carved out by nature alone, to golf courses that are artificially constructed and manipulated in ways that seem inconsistent with their natural setting? Who or what is responsible for the changes in both golf courses and the game of golf? What are the prevalent views of what a golf course should be? If in our exploration we discover that we have taken a turn someplace and we are not where we want to be or should be, we need to reassess where we are and decide how we're going to change. This is by and large the point of this book.

For those of you who are committed to designing and managing golf courses with environmental sensitivity and awareness, I applaud your efforts and encourage you to continue your education and outreach and expand your efforts to protect, conserve, enhance, and manage your golf course habitat. I also encourage you to share your knowledge and vision.

Many people in the golf course industry are trying to do the right thing. They are learning and exploring different approaches to golf course design and management. They are returning to the traditions and heritage of golf, and at the same time incorporating the best that technology has to offer to assist them in keeping the natural image of golf. On the other hand, there still seems to be a consensus that golf courses should be very lush, very green, and very manicured, with low turfgrass height and fast greens. Is this vision driven by golfers themselves? What is the responsibility of golf professionals, superintendents, architects, and the media in determining what golf and golf courses have become and what they will be in the future?

The Value of Golf Courses

As I have said on many occasions, as a type of land development, if golf courses are properly sited, appropriately designed, and well managed, they can represent one of the best types of land development. They can embody all of those positive things that are normally associated with development—such as jobs and taxes and general economic growth—and we must not lose sight of the fact that it is economics that drives the game of golf.

More than that, however, when compared to many other types of development, golf courses can also contribute recreational open space that may include many acres of valuable habitat, such as woods, wetlands, lakes, and streams. In addition to providing habitat for the most common species, they can and often do provide safe refuges for threatened and endangered species of both plants and animals.

The Uniqueness of the Game

Because of its potential value as habitat, the playing field of golf sets it apart from nearly every other sport. It is the only game that has traditionally and historically been played in a "natural" setting, and nature has always played a part in the way the game has been played. The natural landscape and wildlife not only provide a context for the game but are an integral part of it. It is the natural aspect that defines a truly classic golf course.

The rules and ethics of the game are also unique to golf. Like other sports, there are rules that provide structure to the game and guidance to the players, but because of the natural playing field, one may find oneself in the woods or rough, behind a tree, or in some other out-of-play area, with no competitor, no witness, no official nearby to ensure that the rules are followed. Certainly in more formal settings like major tournaments, there are rules officials available as arbiters of the game of golf and there are plenty of witnesses. Nevertheless, there are many stories in golf literature about some of the greatest players of all time and how they assessed themselves penalty strokes during major golf tournaments for infractions only they were privy to and which later resulted in tournament loss. But for the common player, golf has always been considered a game based on honesty. This interrelationship of the rules and ethics of the game of golf and its natural setting are what defines golf's uniqueness.

The Role of Technology

A major impact, however, on what the game of golf has become is technology. With the ever increasing advancement of all sorts of technology, the equipment has changed and so has the playing field. As a result of these changes—and because of television and media coverage—our expectations of what golf "should" be has changed.

Anyone who watches golf on television is aware of the variety of high-tech balls that fly faster and higher, clubs made with new materials that make it easier to hit longer shots, and putters to hit straighter putts. Given the increasing advances in technology, what alternative is there but to make the distance between holes longer, make the greens faster, make the roughs deeper, increase the hazards, change the topography of the green, and grow turfgrass and other vegetation in places it would never grow naturally? No other sport—hockey, basketball, football, tennis, or baseball—has manipulated their playing field to this extent because it would impact the very character of the game itself. And so it has with golf. Over the past seventy years, in order to keep up with the technological changes and make the game more challenging, we have altered the very nature of golf.

With other sports, the playing field is level—a basketball court is the same whether it is in Madison Square Garden or in the local high school gym. The height of the hoop is regulated. The size and shape of the court is regulated. The equipment is standardized. The game is based on skill and talent. The tradition of golf is also based on skill and on the way we face the challenge of the game. However, the equipment, although tested by the USGA, is not "standard," and the playing field is not "regulated." Nevertheless, we have a sense that somehow they should be. We expect that aside from cosmetic variations—variations in landscape and geography—the golf course itself should be like the ones we see on television.

The Role of the Media

Expectations are what we believe "should" be. They are what we imagine, and what we imagine is based in large part on what we see and understand. If the golf course we see on television—the one the professionals play on—sets the standard for what a "perfect" golf course looks like, then somehow we have every right to expect that same vision and that same playing experience. We have the ability to manipulate the playing field, and we have specific ideas about what the playing field should look like. Who decides what the perspective and perception should be? Who drives the decisions about what a golf course is "supposed" to look like and what is their role in making those decisions?

Many of the televised golf tournaments give the impression that all golf courses are, or at least should be, just like televised golf courses. Perhaps not enough is known by the public about the amount of hours, equipment, and number of months or years it took to prepare the course for tournament conditions. Unfortunately, those efforts are not glamorous; they are not newsworthy. So golfers walk away thinking that their golf course should look like the tournament courses, without a clear understanding of how much money and time are required, and how unrealistic their expectations are. Perhaps it is the media's responsibility to focus on the effort it takes to turn a "regular" golf course into a "tournament" golf course.

Perhaps, however, pressure needs to be brought to bear on tournament sponsors so that they will make time to emphasize the "environmental image" of the golf courses. Perhaps it is partly the responsibility of PGA Tour professionals to bring some awareness to the public and to golfers. Perhaps it is golfers themselves who need to better understand the relationship between the heritage and future of golf and the value of golf courses as habitat.

The Role of Golfers

Golf course architects, superintendents, PGA professionals and others associated with the game have said that they design and manage courses to meet the expectations of golfers. The prevailing view is that golfers will accept nothing less than totally "manicured" conditions. To many golfers, that means lush, green, short-mowed, wall-to-wall turfgrass.

Is this **really** what most golfers want? Suppose golfers actually understand that the golf courses they play on are not "showpieces." Suppose "perfection" means different things to different people. Suppose their vision can be modified and their understanding enhanced, and their awareness of the traditions and heritage of golf awakened. Supposing most golfers don't really look for perfection.

The fact of the matter is that most don't. Most golfers merely show up at the course, play the game, and leave for home or work. They don't say much of anything about the course, the traditions of the game, or the environment. However, there always seems to be a small number of vocal golfers with unreasonable expectations about course conditions. These golfers hold strong opinions about how the course should look and how it should be maintained for a variety of reasons. Sometimes their expectations are in response to what they see during televised golf tournaments. Sometimes their expectations are tied to their own financial investment in their course. Whatever the reason, the players who have strong opinions about the manicured, wall-to-wall green look seem to speak up all of the time. This is bad for the game and it makes the life of a golf course superintendent miserable.

We need to encourage the "silent" golfers to more visibly and vocally support the traditions of the game and to make sure that the game of golf retains its historical flavor. We need to help golfers understand the economic impact of reducing the cost of maintenance, reducing adverse impacts on the environment, and increasing the game's accessibility to everyone no matter what their economic status. We need to educate golfers about the environmental issues surrounding the construction and management of golf courses. And we need to applaud those who are supporting the efforts of superintendents and others who are trying to make reasonable decisions about golf that benefits people, wildlife, the environment, and the future of the game.

What Golfers Can Do to Help

The American golf community is dedicated to preserving the game's treasured links to nature. As a result, golf courses are now being developed, designed, and managed more responsibly than ever before. However, we who play the game also have a responsibility to help ensure that golf remains compatible with nature and that our courses are well managed and in harmony with the environment.

As golfers we should:

1. Recognize that golf courses are managed land areas that should complement the natural environment.
2. Respect designated environmentally sensitive areas within the course.
3. Accept the natural limitations and variations of turfgrass plants growing under conditions that protect environmental resources (e.g., brown patches, thinning, loss of color).
4. Support golf course management decisions that protect or enhance the environment, and encourage the development of environmental conservation plans.
5. Support maintenance practices that protect wildlife and natural habitat.
6. Encourage maintenance practices that promote the long-range health of the turf, and support environmental objectives. Such practices include aerification, reduced fertilization, limited play on sensitive turf areas, reduced watering, etc.
7. Commit to long-range conservation efforts (e.g., efficient water use, Integrated Plant Management, etc.) on the golf course and at home.
8. Educate others about the benefits of environmentally responsible golf course management.
9. Support research and education programs that expand our understanding of the relationship between golf and the environment.
10. Take pride in our environmentally responsible courses.

Environmental Principles for Golf Courses in the United States, 1996, The Center for Resource Management.

The Role of Golf Course Architects

Like golfers, there are golf course architects who take into account the traditions of golf as well as the environment when they design golf courses. They do so with a particular eye toward the natural setting and the constraints of the property. Why, then, do some golf course architects design courses that are inconsistent with the landscape and are ultimately very expensive to maintain after they are built? Is it because that's what golfers demand? Is it because they don't know any better? I'd say "no" to both questions.

The golf course design business is very competitive and expensive. Each course designer must have his or her own "trademark." They must make a bold statement that will justify the decision the landowner made to select that course designer to do the job. For some, designing a course that blends into nature does not make a bold enough "television" image. Other architects are dedicated to merely creating a challenging playing surface, rather than merging that part of golf with the heritage of the game.

A course that demonstrates uniqueness and distinction is premised not only on the playing surface, but also on its aesthetic value—the value of the course incorporated in a natural setting. Increasingly, golf course designers are breaking away from an approach that makes neither economic nor environmental sense. They are resisting the temptation to create golf course designs than cannot be maintained without spending great sums of money and time, that use up valuable natural resources, and that are accessible only to the "economically advantaged." Developers, golfers, and the golf industry generally need to support those efforts and applaud those architects whose vision is more consistent with the heritage of the game.

The Role of Golf Course Superintendents

As more golf course architects design golf courses in ways more consistent with the natural landscape, perhaps fewer golf course superintendents will need to use intensive management practices. Some golf course superintendents say their maintenance practices are a direct result of golf course design decisions, which they had nothing to do with, and that they are destined to manage in specific ways as a result. Others will say that their members or players will tolerate nothing less than "perfect" playing conditions. Having visited their courses and spent time with golf course superintendents, I feel there is some truth to these statements.

The fact of the matter is that it is the design of the course that dictates how the course is managed. It is often not the course superintendent who is involved in the selection of turfgrasses to be used on the course, for example. Often the course superintendent has not yet been hired when the irrigation system is selected, designed, and installed. It is not the golf course superintendent who designs course drainage systems that allow for drainpipes to enter directly into course ponds and streams. It is not the golf course superintendent who designs water bodies without littoral shelves and aquatic vegetation. But, the golf course superintendent is the one left behind to work with these decisions.

We need to hire and involve golf course superintendents much sooner in the golf course design and development process than we often do. We need to do everything we can to make environmental education a part of every golf course superintendent's background. Golfers need to support the efforts of their superintendents and become more involved in projects that directly affect the value of the golf course to people, wildlife, the environment, and to the game of golf.

The PGA Professional

We need to distinguish in our own minds between traditional golf and television golf because this is the foundation for many of our unrealistic expectations. The game of golf began as recreation, not as a game that millions of people watch on television. Two-million-dollar purses and million-dollar endorsements and advertisement agreements are a far cry from the way the game was created.

The contribution of PGA professionals is a critical one. They keep golf in the public eye, and this visibility helps to drive the economic engine that runs the game of golf. The public looks up to PGA touring professionals. They are supposed to embody the best that golf can be. Why, then, do PGA touring professionals play on golf courses that are, for the most part, overmanicured, unnaturally green, and artificially immaculate? Are these courses more challenging? Would they be less challenging if they were not as heavily maintained or if more out-of-play areas emphasized a more natural look in contrast to the playing surface? Probably not, but touring professionals will play wherever the tournaments are held.

All professionals want to win, and they want to win by shooting under par—as far under par as they can. This keeps the fans coming to tournaments and turning on the television. This keeps the tournament purses going up, which means more money for the touring professional. So what is the responsibility of the PGA professionals? Since they are in the public eye, they have a responsibility to be aware of the environmental issues facing golf courses. They, along with golfers, architects, and managers, need to be educated and made aware of the value of golf courses as habitat, as naturalized green spaces in increasingly developed areas. They need to bring this into the public arena because they are the ones who can reach the largest audience. PGA professionals need to view themselves as the stewards of the game, the keepers of the history and tradition, the ethics, and the nature that are inherently part of the game.

Doing the Right Thing

Golf and the way that we manage it can represent the best or the worst of humans' relation to the environment. If we retain the traditions of the game both in the playing field and the rules, golf can represent the best that humans have to offer. While we can continue to develop new technologically based equipment, we need to protect the traditions of the rules and the playing field. Even though the rules of golf are often hard to understand, each individual golfer must do his or her best to play by the rules, not because of the rule, but because it is the right thing to do. The same approach must be incorporated into how we build and manage golf courses.

If we choose not to take the path of tradition, the golf course industry will continue to face local opposition to the development of new courses. The public will continue to apply pressure on existing courses by demanding increasing regulations for the use of natural resources and restricting the use of pesticides. The media will continue to show golf courses in the worst possible light. Course development and management expenses will continue to spiral upward. The game will become beyond the reach of the general public and eventually lose its acceptability and popularity. This need not—and should not—happen.

Every golf course architect, superintendent, PGA professional, and golfer shares a responsibility for the future of the game of golf. Because of its relationship with nature, we are caretakers of both the game and the land on which it is played. They are inherently related to each other. This stewardship should be

Ron Dodson, Muirfield, Scotland.
Courtesy of Audubon International.

seated in a respect for the history of the game and the way we play it, and a respect for the land on which it resides. In fostering this attitude, we can expand beyond the course and positively impact the way we live our lives and manage our businesses.

The entire golf industry has the unique potential to be a catalyst for environmental stewardship. Each of us, no matter what our role or responsibility, has the opportunity to follow the lead of the traditions of golf by fostering its history of ethics, civility, and honesty, and by preserving respect for the game, the players, and the nature in which the game is played. Your efforts will be your legacy to future generations. I wish you great success.

12 *Case Studies*

Semiahmoo Golf & Country Club
Blaine, Washington

Semiahmoo has created over 120 acres of natural grassland and wildflowers adjacent to fairways and tees, and restored the understory of trees and shrubs to their natural state to provide habitat for wildlife. Over 300 native trees have also been planted throughout the golf course. Buffer zones of tall grasses and native aquatic plants were established around ponds to ensure water quality and wildlife habitat.

> "Continued involvement in protecting and enhancing the environment and wildlife is the responsibility of us all. It is our hope that this philosophy will lead to greater environmental awareness, both locally as well as on the national level."

Vance Much, Superintendent

Semiahmoo Golf & Country Club, Blaine, Washington.
Courtesy of Audubon International.

Blue Hills Country Club
Kansas City, Missouri

"Blue Hills Country Club is located on 160 acres in a very populated area of southern Kansas City, Missouri. Like many other golf courses, it is completely surrounded by homes. There are three large areas set aside as wildlife sanctuaries with bluebird and purple martin houses, bird feeders, and planted in native grasses and wildflowers. It is not uncommon to see such wildlife as deer, red-tailed hawks, blue egrets, foxes, geese, and once in a while a bobcat. Like many other golf courses, it is a tremendous environmental plus to the surrounding community."

Dave W. Fearis, Superintendent

Blue Hills Country Club, Kansas City, Missouri.
Courtesy of Audubon International.

Old Marsh Golf Club
Palm Beach Gardens, Florida

Old Marsh Golf Club includes 100 acres of turf interspersed with 50 acres of southern mixed forest, approximately 103 acres of protected wetlands, and 35 acres of lakes. In addition, there are 213 private residences on the property, each with buffer zones between them and the protected wetlands. Old Marsh's stewardship activities include protecting the sandhill crane, whose population has doubled on the property in ten years; planting 20 acres of littoral vegetation to protect lakes; landscaping and naturalizing with native plants, and building 20-foot moats as buffer between the golf course and protected wetlands.

"Old Marsh was developed with a long-range vision of protection and enhancement of the natural environment. We are proud of the environmental legacy that has been entrusted to us and we practice sound stewardship in all club, course, housing, and grounds management."

Stephen Ehrbar, Superintendent

Old Marsh Golf Club, Palm Beach Gardens, Florida.
Credit: Alice Semke. Courtesy of Audubon International.

Oregon Golf Club
West Linn, Oregon

"The Oregon Golf club integrates 178 acres of golf course into the natural vegetation and land contours of each hole, with panoramas of Mt. Hood and the Willamette River Valley. The original course design reflects the attention paid to environmental conservation and habitat protection. The Club has focused on maintaining and improving areas for diverse wildlife population inhabiting the course. In partnership with the Oregon State Parks, the Oregon Golf Club is putting in a nature trail through the golf course to the state park, which—in conjunction with the golf course—will also serve as a wildlife corridor."

Russell Vandehey, Superintendent

Oregon Golf Club, West Linn, Oregon.
Courtesy of Audubon International.

Hindman Park Golf Course
Little Rock, Arkansas

"Hindman Park, a municipal golf course, is interspersed with woodlands, native grasses, wildflowers, and creeks—habitat that provides refuge for a diverse wildlife population. The outside area of the golf course is left in its natural condition in order to provide not only a quality golfing experience, but also a beautiful nature area that is enjoyed by bird-watchers and many others who just like to take a walk in the woods.

"In 1996, we planted over 100 fruit trees plus some shrubs and ornamentals that are fruit- and berry-bearing. This will provide cover, but more importantly, a natural food source for our birds and wildlife. These trees and shrubs will not be maintained in any way, but left to grow and prosper at nature's will. We intend to continue this for at least the next five years, giving us about 500 new trees of many different varieties. Think what *that* will be like 20 years from now. We are not only stewards of this land and all that is on it, but caretakers for the next generation, and they in their turn must leave it better again."

John Miller, Jr., Superintendent

Hindman Park Golf Course, Little Rock, Arkansas.
Credit: Kathy Hinson. Courtesy of Audubon International.

Prairie Dunes Country Club
Hutchinson, Kansas

Prairie Dunes Country Club is a private 18-hole golf course built in the 1930s on nearly 200 acres of sand dunes. The course is laced with tall-grass prairie, and surrounded by rangeland, housing development, and remnant prairie, and serves as an integral part of the larger greenspace corridor in the area. Seventy-five percent of the property is maintained as natural habitat, and Prairie Dunes serves as a research site for Tabor College.

"Imagine being on hand for every sunrise at the world's most beautiful property, and you can easily understand a superintendent's enthusiasm for and appreciation of the successful marriage between golf and nature. Conservation and environmental enhancement are a natural offspring of the rapport superintendents have with the earth."

P. Stan George, Superintendent

Prairie Dunes Country Club, Hutchinson, Kansas.
Courtesy of Audubon International.

Hole-in-the-Wall Golf Club
Naples, Florida

Opened in 1952, Hole-in-the-Wall Golf Club was Naples' first private 18-hole golf course. The club includes 50 acres of cypress swamps and woods and 10 acres of lakes, which provide for a diversity of wildlife, such as gray fox, bobcat, Cypress fox squirrel, wood stork, and osprey. Hole-in-the-Wall serves as a research site for the University of Florida for a Cypress Fox Squirrel study. Waterway shorelines are landscaped with extensive aquatic plantings to increase habitat and decrease erosion.

"We have been amazed at how little cost and little effort there can be in a program making a golf course more compatible with the environment."

Fred Yarrington, Environmental Committee Chair

Hole-in-the-Wall Golf Club, Naples, Florida.
Courtesy of Audubon International.

Village Links of Glen Ellyn
Glen Ellyn, Illinois

When the golf course was built in 1967, virtually all of the original land was disturbed and reshaped into the existing property. The property consisted of turf-grass that was mowed from fence line to fence line. Only a couple of dozen preconstruction trees remained. In the late 1980s a program to "naturalize" the golf course was initiated. More than 50% of the shorelines have unmowed buffers. Some areas were planted with sapling trees and shrubs, others with wildflowers. Fairway coverage was reduced from 62 to 42 acres and over 40 acres of the course were restored to native grassland and prairie habitat. The course also planted six acres of native trees to create woodland. To learn more about the local ecosystem, the Village Links of Glen Ellyn staff worked closely with several groups in the greater Chicago area—including the Morton Arboretum, Illinois Department of Conservation, Max McGraw Wildlife Foundation, and Wild Birds Unlimited—who have all provided valuable information and advice about local wildlife and native plants.

"Our goal is to try to enhance wildlife habitat so our golf course can provide ongoing support for a diverse variety of wildlife, especially species that are not currently present on the property due to the loss of native habitat."

Chris Pekarek, Assistant Superintendent

Village Links of Glen Ellyn, Glen Ellyn, Illinois.
Courtesy of Audubon International.

The Ivanhoe Club
Ivanhoe, Illinois

Located in Lake County, Illinois, in a relatively undeveloped section of the Chicagoland metropolitan area, the 250-acre golf course was developed as part of a 700-acre housing development. The Ivanhoe Club has evolved into a 27-hole golf course with three distinct habitat themes—the Marsh, the Prairie, and the Forest—for each nine holes. Native grasses and wildflowers are used in the prairie area, water plants in the marsh area, and native hardwood trees in the forest area. No-mow areas have been established in large tree understory areas and along prairie habitat. Ivanhoe's Integrated Pest Management program included reducing insecticide use by 90% in 1995, and in 1996 no insecticides were used on any turf at the Ivanhoe Club.

"I like the challenge of saving money, saving the work, and saving the insects for the birds. To me it's kind of like a game, and the less I mess with insecticides, the better off our golf course is going to be. We make a very simple statement: we give more life than we take. Our golf course is a sanctuary for wildlife and we make every effort to avoid interfering with the basic food chain."

Peter Leuzinger, Superintendent

Ivanhoe Club, Ivanhoe, Illinois.
Courtesy of Aubudon International.

Schuyler Meadows Club
Loudonville, New York

Located just outside of Albany, New York, tucked in a primarily residential area, Schuyler Meadows Club is a 200-acre golf course that was established in 1926. Once farmland, the habitat now includes 110 acres of woods, 8 acres of tall grass meadows, a 3-acre pond, and 3 acres of wetlands. It has two small streams that run year-round and one small stream that periodically dries up in the summer. Schuyler Meadows' neighbor is Siena College, which—in partnership with Audubon International—is currently conducting wildlife research, including manipulating and monitoring maintenance practices on the course to assess the impact on wildlife species.

"Naturalizing has added a new challenge to our golfers' game, and they're happy. We have much more habitat for our wildlife, and they're happy. We spend a lot less time mowing rough, and I'm very happy. Talk about a win-win situation."

Peter Salinetti, Superintendent and General Manager

Schuyler Meadows Club, Loudonville, New York.

Glendale Country Club
Bellevue, Washington

Glendale Country Club is a private member club located in the city of Bellevue, Washington. The course features tree-lined fairways with sloped terrain falling toward two salmon spawning creeks. It is bordered to the south by a city park, to the north by a major thoroughfare, and private residences on the other sides. Kelsey Creek had been mowed and maintained as part of the playing areas until 1990, when it was allowed to revert to a natural state. Sections were planted with willows and schoolchildren were allowed to release salmon fingerlings raised as part of a class project. In conjunction with the City of Bellevue's Storm and Surface Water Utility's "Stream Team," 7,500 salmon fingerlings were released into Kelsey Creek after it was restored with new vegetation to reduce erosion.

Steve Kealy, Superintendent

Glendale Country Club, Bellevue, Washington.
Courtesy of Audubon International.

Country Club of Florida
Village of Golf, Florida

One of the most visible enhancement projects at the Country Club of Florida is the Wildlife Corridor. Consisting of four acres, this area does not affect play of the course, but certainly makes the golf course appealing for wildlife. The intent of the corridor is to connect two large bodies of water that are out of play, and give wildlife room to migrate from one water source to the other. Within this corridor is a wetland, an island with purple martin houses and duck boxes, brush piles, bird feeders, and hundreds of native plants that provide food sources for wildlife.

"At the Country Club of Florida it has always been our goal to provide the best possible course conditions and quality for our members. Our members are proud of the efforts that have been under way to enhance the golf course for wildlife."

Jeff Klontz, Superintendent

Country Club of Florida, Village of Golf, Florida.
Courtesy of Audubon International.

Quail Run Golf Course
Lupine, Oregon

Before its establishment in 1991, Quail Run Golf Course was a cattle ranch. It is a true mountain course located on the east side of the Cascades. The course is located in a relatively rural area, so urban pressures on wildlife aren't as great as compared with a big city. Nevertheless, a nest box program was incorporated that now boasts 40 nest boxes on the property and a bird inventory list of approximately 105 birds—including a breeding pair of great gray owls that have been in residence for more than five years.

"We feel we are managing the golf course with the big picture in mind. It functions as a healthy, viable ecosystem and still provides for a quality golfing experience. I used to be a forester and managed timberland. I understand the importance of taking care of the land and think that taking an ecosystem approach to managing the golf course is the right thing to do."

Jim Peterson, Superintendent

Quail Run Golf Course, Lupine, Oregon.
Courtesy of Audubon International.

TPC at the Canyons
Las Vegas, Nevada

"The design and routing of the golf course through the native desert has almost naturally created or preserved wildlife corridors. When designing a course in the desert, nature has already created natural washes and canyons that must be left intact to allow for rainfall to move through the property. Large corridors have been naturally created and have been undisturbed by minimizing the amount of turf used to create the course. We really do nothing out of the ordinary to create corridors, as they have occurred naturally and were left that way by the architect.

"Our facility serves as a home for one endangered species, the desert tortoise. The Department of Wildlife has encouraged us to let the tortoise alone, and do nothing except keep its habitat undisturbed, which is how we treat the desert. In our opinion the desert should be left alone. It has evolved to where it is today naturally and we wish to continue that evolution."

Kim Wood, Superintendent

TPC at the Canyons, Las Vegas, Nevada.
Courtesy of Audubon International.

Resources

There are a variety of professional organizations that provide support, education, and additional resources for golf professionals, especially regarding turfgrass and environmental issues relative to golf courses.

The **United States Golf Association** has supported environmental research through its Turfgrass and Environmental Research Committee. Additionally, the USGA's Green Section provides the Turf Advisory Service, which includes onsite course visits, consultation, and education.

The **National Fish and Wildlife Foundation** and the **United States Golf Association** have developed a partnership called **Wildlife Links**. Wildlife Links is a cooperative program that funds research, management, and education projects that will help golf courses become an important part of the conservation landscape. The United States Golf Association is providing $200,000 annually to fund the grants through the Wildlife Links program, which is managed by the National Fish and Wildlife Foundation.

Ongoing training by the **Golf Course Superintendents Association of America (GCSAA)** is available to golf course superintendents with access to information about environmentally sound management of fertilizers, water, and pesticides.

Members of the **American Society of Golf Course Architects (ASGCA)** work cooperatively with local zoning and regulatory agencies to promote and implement innovative design solutions to protect the environment.

Audubon International is a not-for-profit environmental organization dedicated to sustainable resource management and environmental conservation through education, research, and public involvement. Audubon International promotes ecologically sound land management, habitat restoration, and natural resource conservation.

Audubon International manages two membership programs: the **Audubon Cooperative Sanctuary System (ACSS),** and **The Audubon Signature Program**. The ACSS is an educational program for existing facilities that provides information, guidance, and support to individuals, schools, businesses, and golf courses to encourage people to become personally involved in managing habitat for both people and wildlife. The **Audubon Signature Program** is a conservation assistance program designed to work with projects in the planning stages of development to ensure that land development occurs with a primary focus on sustainable natural resource management.

   The Audubon Cooperative Sanctuary Program for Golf Courses offers an environmental education program for golf course managers that provides a framework for and information about developing and implementing an environmental plan that includes a variety of conservation and enhancement activities, and provides recognition for professionals who demonstrate an active commitment to an environmental approach to managing their golf course.

Golf

American Society of Golf Course Architects
221 N. LaSalle St.
Chicago, IL 60601-1520
(321) 372-7090
http://www.golfdesign.org

Golf Course Superintendents Association of America
1421 Research Park Drive
Lawrence, KS 66049
(913) 832-2240
http://www.gcsaa.org

United States Golf Association
Golf House
P.O. Box 708
Far Hills, NJ 07931
(908) 234-2300
http://www.usga.org

Environmental

Audubon International
46 Rarick Road
Selkirk, NY 12158
(518) 767-9051
http://www.audubonintl.org

Cornell Lab of Ornithology
159 Sapsucker Woods Road
Ithaca, NY 14850
(607) 254-2442
http://www.ornith.cornell.edu

National Fish and Wildlife Foundation
1120 Connecticut Ave., NW, Suite 900
Washington, DC 20036
(202) 857-20036
http://www.nfwf.org

Reference and Reading List

Benton and Werner. *Field Biology and Ecology*. McGraw Hill Publishers.

California Center for Wildlife, with Diana Landau and Shelley Stump. *Living with Wildlife: How to Enjoy, Cope with, and Protect North America's Wild Creatures Around Your Home and Theirs*. Sierra Club Books. 1994

Collins, Henry Hill Jr. *Complete Field Guide to American Wildlife*. Harper and Row Publishers. 1959.

Cox, Jeff. *Landscaping with Nature*. Rodale Press Publishers. 1991.

Decker, Daniel J. and John W. Kelley. *Enhancement of Wildlife on Private Lands*. Cornell Cooperative Extension Publication. Undated.

Doak, Tom. *Anatomy of a Golf Course*. Lyons and Burford Publishers. 1992.

Ehrlich, Paul R., David S. Dobkin, and Darryl Wheye. *The Birder's Handbook: A Field Guide to the Natural History of North American Birds*. Simon and Schuster, Inc. 1998

Golf Course Superintendents Association of America. *Golf Course Design*. Undated.

Harker, et al. *Landscape Restoration Handbook*. United States Golf Association. Lewis Publishers. 1993.

Hunter, Malcom L. *Fundamentals of Conservation Biology*. Blackwell Science Publishers. 1986.

Jarrett, Tom. *St. Andrews Golf Links—The First Six Hundred Years*. Mainstream Publishing. 1995.

Mackay, Jean. *A Guide to Environmental Stewardship on the Golf Course*. Audubon International 1996.

McCord, Robert R. *Golf: An Album of Its History*. Burford Books, Inc. 1998.

Meriles, Bill. *Attracting Backyard Wildlife*. Voyageur Press Publishers. 1989.

Minnesota Department of Natural Resources, *Landscaping for Wildlife*. 1986.

Nebraska Game and Parks Commission. *Wildlife Habitat Improvement Guide*. 1991.

Price, Robert. *Scotland's Golf Courses.* The Mercat Press Publishers. 1992.

Smith, Robert Leo. *Ecology and Field Biology.* Harper and Row Publishers. 1980.

Stokes, Don & Lillian. *The Bird Feeder Book.* Little, Brown & Co. 1987.

Wilson, Edward O. *The Diversity of Life.* W.W. Norton and Company Publishers. 1992.

Field Guides

Field guides will help you identify a variety of plant and wildlife species. They are often the best source of increasing your education and awareness of the property where you live and work. Series of field guides include such species identification as plants, trees and shrubs, birds, mammals, reptiles and amphibians, wildflowers, and insects. Visit your local library or bookstore and browse for the ones that seem to best suit your needs and interests.

American Bird Conservancy: All the Birds of North America
Audubon Society Field Guides
Golden Field Guides
National Geographic Society Field Guides
Peterson Field Guides
Stokes Field Guides
Thayer Birds of North America (CD-Rom), Version 2.5
        Thayer Birding Software, 1998

Appendix

WILDLIFE MANAGEMENT AND NATURAL RESOURCE AGENCIES AND ORGANIZATIONS

Below are state governmental agencies that might provide some assistance regarding wildlife and natural resource management questions or direction regarding other organizations or agencies that might be appropriate to contact. Also listed are Cooperative Fish and Wildlife Research Units (a partnership among the Biological Resources Division of the U.S. Department of the Interior, state fish and game agencies, a host university, and the Wildlife Management Institute).

ALABAMA

Alabama Cooperative Fish & Wildlife Research Unit
331 Funchess Hall
Auburn University
Auburn, AL 36849
(334) 844-4796

Alabama Department of Conservation & Natural Resources
64 N. Union, Street
Montgomery, AL 36130
(334) 242-3486

ALASKA

Alaska Cooperative Fish & Wildlife Research Unit
P.O. Box 757020
University of Alaska–Fairbanks
Fairbanks, AK 99775-7020
(907) 474-7661

Department of Environmental Conservation
410 Willoughby Avenue
Juneau, AK 99801-1795
(907) 465-5000

Department of Fish & Game
P.O. Box 25526
Juneau, AK 99802
(907) 465-4100

Department of Natural Resources
400 Willoughby Avenue, 5th Floor
Juneau, AK 99081
(907) 465-2400

ARIZONA

Arizona Cooperative Fish and Wildlife Research Unit
Biological Sciences East, Room 104
University of Arizona
Tucson, AZ 85721
(520) 621-1959

Arizona Game and Fish Department
2221 W. Greenway Road
Phoenix, AZ 85023-4312
(602) 942-3000

ARKANSAS

Arkansas Cooperative Research Unit
Biological Sciences, SCEN 617
University of Arkansas
Fayetteville, AR 72701
(501) 575-6709

CALIFORNIA

Department of Conservation
801 K Street, 24th Floor
Sacramento, CA 95814
(916) 322-1080

Department of Fish & Game
1416 9th Street
Sacramento, CA 95814
(916) 653-7664

COLORADO

Colorado Cooperative Fish & Wildlife Research Unit
201 Wagar Building
Department of Fishery and Wildlife Biology
Colorado State University
Ft. Collins, CO 80523-1484
(907) 491-5396

Department of Natural Resources
1313 Sherman, Room 718
Denver, CO 80203
(303) 866-3311

CONNECTICUT

University of Connecticut Cooperative Extension
College of Agriculture and Natural Resources
Box U-66
1376 Storrs Road
University of Connecticut
Storrs, CT 06269-4066
(203) 486-2840

DELAWARE

Department of Natural Resources & Environmental Control
89 Kings Highway
P.O. Box 1401
Dover, DE 19903
(302) 739-4506

FLORIDA

Florida Cooperative Fish & Wildlife Research Unit
117 Newins-Ziegler Hall
P.O. Box 110450
University of Florida
Gainesville, FL 32611-0450
(904) 392-1861

GEORGIA

Department of Natural Resources
Wildlife Resources Division
2070 U.S. Highway 278, SE
Social Circle, GA 30279
(404) 918-6401

Georgia Cooperative Fish & Wildlife Research Unit
Warnell School of Forest Resources
University of Georgia
Athens, GA 30602-2152
(707) 546-2234

HAWAII

Department of Land & Natural Resources
Box 621
Honolulu, HI 96809
(808) 587-0400

Hawaii Cooperative Fishery Research Unit
2538 The Mall
University of Hawaii
Honolulu, HI 96822-2279
(808) 956-8350

IDAHO

Fish & Game Department
600 S. Walnut
Box 25
Boise, ID 83707
(208) 334-2114

Idaho Cooperative Fish & Wildlife Research Unit
College of Forestry
Wildlife and Range Sciences
University of Idaho
Moscow, ID 83844-1136
(208) 885-6336

ILLINOIS

Illinois Department of Natural Resources
524 S. 2nd Street, Room 400 LTP
Springfield, IL 62701-1787
(217) 785-0067

INDIANA

Indiana Department of Natural Resources
402 W. Washington Street, Room W255B
Indianapolis, IN 46204
(317) 232-4200

IOWA

Department of Natural Resources
E. 9th & Grand Avenue, Wallace Building
Des Moines, IA 50319-0034
(515) 281-5145

Iowa Cooperative Fish & Wildlife Research Unit
Science Hall II Animal Ecology Department
Iowa State University
Ames, IA 50011-3221
(515) 294-3056

KANSAS

Kansas Department of Wildlife & Parks
900 SW Jackson Street, Suite 502
Topeka, KS 66612-1233
(913) 296-2281

Kansas Cooperative Fish & Wildlife Research Unit
205 Leasure Hall
Kansas State University
Manhattan, KS 66506-3501

KENTUCKY

Department of Fish & Wildlife Resources
#1 Game Farm Road
Frankfort, KY 40601
(502) 564-6508

LOUISIANA

Louisiana Department of Wildlife & Fisheries
P.O. Box 98000
Baton Rouge, LA 70898-9000
(504) 765-2800

Louisiana Cooperative Fish & Wildlife Research Unit
School of Forestry, Wildlife & Fisheries
Louisiana State University
Baton Rouge, LA 70803-4209
(504) 388-4179

Grambling Cooperative Wildlife Research Project
Department of Biological Sciences
Box 4290
Grambling State University
Grambling, LA 71245-4290
(318) 274-2499

MAINE

Department of Inland Fisheries and Wildlife
284 State Street, Station #41
Augusta, ME 04333
(207) 287-2766

Maine Cooperative Fish & Wildlife Research Unit
5755 Nutting Hall, Room 206
University of Maine
Orono, ME 04469-5755
(207) 581-2901

MARYLAND

Maryland Department of Natural Resources
Wildlife Division
Tawes State Office Building
Annapolis, MD 21401
(410) 974-3105

Maryland Cooperative Fish & Wildlife Research Unit
University of Maryland (Eastern Shore)
Trigg Hall
Princess Anne, MD 21853-1299
(410) 651-7663

MASSACHUSETTS

Office of Environmental Affairs
Department of Fisheries, Wildlife & Environmental Law Enforcement
100 Cambridge Street, Room 1901
Boston, MA 02202
(617) 727-3155

Massachusetts Cooperative Fish & Wildlife Research Unit
Forestry and Wildlife
Holdsworth Natural Resource Center
University of Massachusetts
Box 34220
Amherst, MA 01003-4220
(413) 545-0080

MICHIGAN

Department of Natural Resources
Box 30028
Lansing, MI 48909
(517) 335-4623

MINNESOTA

Department of Natural Resources
500 Lafayette Road
St. Paul, MN 55155-4001
(612) 297-4946

Minnesota Cooperative Fish & Wildlife Research Unit
Department of Fisheries and Wildlife
College of Natural Resources
200 Hodson Hall
University of Minnesota
St. Paul, MN 55108-6124
(612) 624-3421

MISSISSIPPI

Department of Wildlife, Fisheries, and Parks
P.O. Box 451
Jackson, MS 39205
(601) 362-9212)

Mississippi Cooperative Fish & Wildlife Research Unit
P.O. Drawer BX
Mississippi State University
Mississippi State, MS 39762
(601) 325-2643

MISSOURI

Missouri Cooperative Fish & Wildlife Research
112 Stephens Hall
University of Missouri
Columbia, MO 65211
(573) 882-3634

Department of Natural Resources
P.O. Box 176
Jefferson City, MO 65102
(314) 751-4422

MONTANA

Department of Natural Resources & Conservation
1625 11th Avenue
P.O. Box 201601
Helena, MT 59620-2301
(406) 444-2074

Montana Cooperative Fishery Research Unit
Department of Biology
Montana State University
Bozeman, MT 57917-0346
(406) 994-4549

Montana Cooperative Wildlife Research Unit
Natural Science 205
University of Montana
Missoula, MT 59812-1120
(406) 243-5372

NEBRASKA

Nebraska Natural Resources Commission
301 Centennial Mall S., 4th Floor
P.O. Box 94876
Lincoln, NE 68509-4876
(402) 471-2081

NEVADA

Department of Conservation & Natural Resources
Capitol Complex
123 W. Nye Lane
Carson City, NV 89710
(702) 687-4360

NEW HAMPSHIRE

Fish & Game Department
2 Hazen Drive
Concord, NH 03302-0095
(603) 271-3503

NEW JERSEY

Department of Environmental Protection
Division of Fish, Game, & Wildlife
CN 400
Trenton, NJ 08625-0400
(609) 292-2965

NEW MEXICO

Energy, Minerals, and Natural Resources Department
2040 Pacheco Street
Santa Fe, NM 87505
(505) 827-5950

New Mexico Cooperative Fish & Wildlife Research Unit
Department of Fisheries and Wildlife Sciences
P.O. Box 30003, Department 4901
New Mexico University
Las Cruces, NM 88003-0003

NEW YORK

Department of Environmental Conservation
Division of Fish & Wildlife
50 Wolf Road
Albany, NY 12233
(518) 457-5690

New York Cooperative Fish & Wildlife Research Unit
Fernow Hall
Cornell University
Ithaca, NY 14853-0188

NORTH CAROLINA

Wildlife Resources Commission
512 N. Salisbury Street, Archdale Building
Raleigh, NC 27604-1188
(919) 733-3391

Cooperative Fish & Wildlife Research Unit
Box 7617, 4105 Gardner Hall
North Carolina State University
Raleigh, NC 27695
(919) 515-2631

NORTH DAKOTA

Institute for Ecological Studies
P.O. Box 7110
University of North Dakota
Grand Forks, ND 58202
(701) 777-4215

State Game & Fish Department
100 N. Bismarck Expressway
Bismarck, ND 58501
(701) 328-6300

OHIO

Department of Natural Resources
Fountain Square
Columbus, OH 43224
(614) 265-6565

Ohio Cooperative Fish & Wildlife Research Unit
1735 Neil Avenue
Ohio State University
Columbus, OH 43210-1293

OKLAHOMA

Department of Wildlife Conservation
1801 N. Lincoln
P.O. Box 53465
Oklahoma City, OK 73152
(405) 521-3851

Oklahoma Cooperative Fish & Wildlife Research Unit
404 Life Sciences West
Oklahoma State University
Stillwater, OK 74078-0611

OREGON

Department of Fish & Wildlife
2501 SW 1st Avenue
P.O. Box 59
Portland, OR 97207
(503) 229-5410

Oregon Cooperative Fish & Wildlife Research Unit
104 Nash Hall
Oregon State University
Corvallis, OR 97331-3803
(541) 737-1938

PENNSYLVANIA

Department of Environmental Resources
Public Liaison Office
16th Floor, MSSOB
P.O. Box 2063
Harrisburg, PA 17105-2063
(717) 783-2300

Pennsylvania Cooperative Fish & Wildlife Research Unit
113 Merkle Building
Pennsylvania State University
University Park, PA 16802-1100

RHODE ISLAND

Department of Environmental Management
235 Promenate Street
Providence, RI 02908
(401) 277-2080

SOUTH CAROLINA

Department of Natural Resources
Rembert C. Dennis Building
P.O. Box 167
Columbia, SC 29202
(803) 734-3888

South Carolina Cooperative Fish & Wildlife Research Unit
308 Lehotsky Hall
Department of Aquaculture, Fisheries & Wildlife
Clemson University
Clemson, SC 29634-0362
(803) 656-0168

SOUTH DAKOTA

Department of Environment and Natural Resources
523 E. Capitol, Joe Foss Office Building
Pierre, SD 57501
(605) 773-3151

South Dakota Cooperative Fish & Wildlife Research Unit
Department of Wildlife & Fisheries Sciences
South Dakota State University
P.O. Box 2140B
Brookings, SD 57007-1696
(605) 688-6121

TENNESSEE

Department of Environment and Conservation
401 Church Street
Nashville, TN 37243
(615) 532-0109

Wildlife Resources Agency
P.O. Box 40747, Ellington Agricultural Center
Nashville, TN 37204
(615) 781-6500

Tennessee Cooperative Fishery Research Unit
Box 5114
Tennessee Tech University
Cookesville, TN 38505-0001
(615) 372-3032

TEXAS

Parks and Wildlife Department
4200 Smith School Road
Austin, TX 78744
(512) 389-4800

Texas Cooperative Fish & Wildlife Research Unit
Department of Range & Wildlife Management
Texas Tech University
Lubbock, TX 79409-2125
(806) 742-2851

UTAH

State Department of Natural Resources
1636 W. N. Temple, Suite 316
Salt Lake City, UT 84116-3193
(801) 538-7200

Utah Cooperative Fish & Wildlife Research Unit
Utah State University
Logan, UT 84322-5210
(801) 797-2509

VERMONT

Agency of Natural Resources
103 S. Main Street
Waterbury, VT 05671
(802) 241-3600

Vermont Cooperative Fish & Wildlife Research Unit
School of Natural Resources
University of Vermont
Burlington, VT 05405-0088
(802) 656-3011

VIRGINIA

Department of Game and Inland Fisheries
4010 W. Broad Street
P.O. Box 11104
Richmond, VA 23230
(804) 367-1000

Virginia Cooperative Fish & Wildlife Research Unit
106 Cheatham Hall
Virginia Polytechnic Institute and State University
Blacksburg, VA 24061
(540) 231-5927

WASHINGTON

Department of Fish & Wildlife
600 Capitol Way, North
Olympia, WA 98501-1091
(206) 902-2200

Department of Natural Resources
P.O. Box 47001
Olympia, WA 98504-7001
(206) 902-1000

Washington Cooperative Fish & Wildlife Research Unit
School of Fisheries WH
University of Washington
Seattle, WA 98195-7980
(206) 543-6475

WEST VIRGINIA

Division of Natural Resources
1900 Kanawha Boulevard, East
Charleston, WV 25305
(304) 558-2754

West Virginia Cooperative Fish & Wildlife Research Unit
Division of Forestry
College of Agriculture and Forestry
P.O. Box 6125
West Virginia University
Morgantown, WV 26506-3794
(304) 293-3794

WISCONSIN

Department of Natural Resources
Box 7921
Madison, WI 53707
(608) 266-2621

Wisconsin Cooperative Wildlife Research Unit
226 Russell Laboratories
University of Wisconsin
Madison, WI 53706-1598
(608) 263-4519

Wisconsin Cooperative Fishery Research Unit
College of Natural Resources
University of Wisconsin
Stevens Point, WI 54481-3897
(715) 346-2178

WYOMING

Game & Fish Department
5400 Bishop Boulevard
Cheyenne, WY 82006
(307) 777-4600

Wyoming Cooperative Fish & Wildlife Research Unit
Biological Science Building, Room 419
Box 3166
University of Wyoming
Laramie, WY 82071-3166
(307) 766-5415

ENVIRONMENTAL PROTECTION AGENCY

The focus of the EPA is on problems of air and water pollution, management of solid and hazardous wastes, cleanup of hazardous wastes under SuperFund, and regulation of pesticides and toxic substances. Functions include setting and enforcing environmental standards; conducting research on the causes, effects, and control of environmental problems, and assisting states and local governments.

Administrator
Environmental Protection Agency
201 M Street, NW
Washington, DC 20406

Region 1
Connecticut, Maine, New Hampshire, Rhode Island, Vermont

Regional Director
EPA: Region 1
1 Congress Street, JFK Building
Boston, MA 02203-0001
(617) 565-3400

Region 2
New Jersey, New York, Puerto Rico, Virgin Islands

Regional Director
EPA: Region 2
290 Broadway
New York, NY 10007-1866
(212) 637-5000

Region 3
Delaware, Maryland, Pennsylvania, Virginia, West Virginia, District of Columbia

Regional Director
EPA: Region 3
841 Chestnut Building
Philadelphia, PA 19107
(215) 597-9814

Region 4
Alabama, Florida, Georgia, Kentucky, Mississippi, North Carolina, South
     Carolina, Tennessee

Regional Director
EPA: Region 4
100 Alabama Street, SW
Atlanta, GA 30303
(404) 347-4728

Region 5
Illinois, Indiana, Michigan, Minnesota, Ohio, Wisconsin

Regional Director
EPA: Region 5
77 West Jackson Boulevard
Chicago, IL 60604-3507
(312) 886-3000

Region 6
Arkansas, Louisiana, New Mexico, Oklahoma, Texas

Regional Director
EPA: Region 6
1445 Ross Avenue, Fountain Place, #1200
Dallas, TX 75202-2733
(214) 665-2100

Region 7
Iowa, Kansas, Missouri, Nebraska

Regional Director
EPA: Region 7
726 Minnesota Avenue
Kansas City, KS 66101
(913) 551-7006

Region 8
Colorado, Montana, North Dakota, South Dakota, Utah, Wyoming

Regional Director
EPA: Region 8
999 18th Street, Suite 500
Denver, CO 80202-2466
(303) 293-1616

Region 9
Arizona, California, Hawaii, Nevada, Guam, American Samoa

Regional Director
EPA: Region 9
75 Hawthorne Street
San Francisco, CA 94105
(415) 744-1001

Region 10
Alaska, Idaho, Oregon, Washington

Regional Director
EPA: Regional 10
1200 Sixth Avenue
Seattle, WA 98101
(206) 553-1234

U.S. ARMY CORPS OF ENGINEERS

The Corps of Engineers provides quality, responsive engineering, and environmental services to the nation. The Corps plans, designs, builds, and operates water resources and other civil works projects. The Corps has a diverse workforce of biologists, geologists, hydrologists, and natural resource managers.

Chief of Engineers
Army Department Corps of Engineers
20 Massachusetts Avenue, NW
Washington, DC 20314-1000
(202) 761-0001

U.S. Army Engineer Division
South Atlantic Division
77 Forsyth Street, SW, Room 322
Atlanta, GA 30303-3490
(404) 331-7444

U.S. Army Engineer Division
Pacific Ocean Division
Building 230
Ft. Shafter, HI 96858-5440
(808) 438-9862

U.S. Army Engineer Division
Mississippi Valley
P.O. Box 80
Vicksburg, MS 39181-0080
(601) 634-5757

U.S. Army Engineer Division
North Atlantic Division
90 Church Street
New York, NY 10007-2979
(212) 264-7500

U.S. Army Engineer Division
Great Lakes & Ohio River
111 North Canal Street
Chicago, IL 60606-7205
(312) 353-6317

U.S. Army Engineer Division
Northwestern Regional Headquarters
12565 West Center Road
Omaha, NE 68144
(401) 697-2600

U.S. Army Engineer Division
South Pacific Division
333 Market Street
San Francisco, CA 94105-2195
(415) 977-8200

U.S. Army Engineer Division
Southwestern Division
1114 Commerce Street, Room 413A
Dallas, TX 75242-0216
(214) 767-2334

REGIONAL U.S. FISH & WILDLIFE SERVICE

The U.S. Fish and Wildlife Service is the federal agency established to protect and conserve the nation's wildlife resources. The agency contains a storehouse of information on wildlife, including birds, butterflies, and threatened or endangered wildlife. The Service is geographically organized in seven regions. Staff at regional offices can provide information and technical assistance about habitat needs of wildlife in your area and native plants needed to support them. The website for the Fish and Wildlife Service is: http://www.fws.gov

Region 1: Pacific Regional Office
Oregon, Washington, Idaho, Nevada, California, Hawaii, Pacific Islands

Regional Director
Eastside Federal Complex
911 NE 11th Avenue
Portland, OR 97232-4181
(503) 231-6118

Region 2: Southwest Regional Office
Texas, New Mexico, Oklahoma, Arizona

Regional Director
U.S. Fish & Wildlife Service-Region 2
500 Gold Avenue, SW, Room #3018
Albuquerque, NM 87102
(505) 248-6282

Region 3: Great Lakes-Big Rivers Regional Office
Minnesota, Wisconsin, Michigan, Iowa, Illinois, Ohio, Indiana, Missouri

Regional Director
U.S. Fish & Wildlife Service-Region 3
1 Federal Drive, FBW Federal Building
Fort Snelling, MN 55111
(612) 725-3563

Region 4: Southeast Regional Office
Louisiana, Alabama, Arkansas, Tennessee, Kentucky, Georgia, Mississippi,
    Florida, Georgia, South Carolina, North Carolina

Regional Director
U.S. Fish & Wildlife Service-Region 4
1875 Century Boulevard
Atlanta, GA 30345
(404) 679-4000

Region 5: Northeast Regional Office
Connecticut, Delaware, Maine, Maryland, Massachusetts, New Hampshire, New
    York, New Jersey, Pennsylvania, Rhode Island, Vermont, Virginia, West
    Virginia

Regional Director
U.S. Fish & Wildlife Service-Region 5
300 Westgate Center Drive
Hadley, MA 01035
(413) 253-8300

Region 6: Mountain-Prairie Regional Office
Colorado, Kansas, Montana, North Dakota, Nebraska, South Dakota, Utah,
    Wyoming

Regional Director
U.S. Fish & Wildlife Service-Region 6
P.O. Box 25486
Denver, CO 80225
(303) 236-7920

Region 7: Alaska Regional Office
Alaska

Regional Director
U.S. Fish & Wildlife Service-Region 7
1011 E. Tudor Road, Room 229
Anchorage, AK 99503
(907) 786-3542

Index